ANALYZING
QUALITATIVE DATA

ANALYZING QUALITATIVE DATA

Introductory Log-Linear Analysis for Behavioral Research

John J. Kennedy

The Ohio State University

PRAEGER SPECIAL STUDIES • PRAEGER SCIENTIFIC

Library of Congress Cataloging in Publication Data

Kennedy, John J.
  Analyzing qualitative data.

  Includes bibliographical references and indexes.
  1. Log-linear models. 2. Psychometrics. 3. Psychological research. 4. Educational research. I. Title.
BF39.K457   1983   150'.1'5195   82-18040
ISBN 0-03-060422-2

Published in 1983 by Praeger Publishers
CBS Educational and Professional Publishing
a Division of CBS Inc.
521 Fifth Avenue, New York, New York 10175 U.S.A.

3456789 052   987654321

Printed in the United States of America

To JUM C. NUNNALLY

a fine teacher and excellent author

Preface

This book has been written for behavioral researchers and students of behavioral research. Its purpose is to introduce and explicate the basic features of log-linear contingency table analysis, a relatively new generalized approach to the analysis of qualitative/categorical data. As will be shown, the principal value of log-linear methods is that they can accommodate a simultaneous analysis of two *or more* qualitative/categorical variables and do so in a manner similar to that seen in the analysis of variance and multiple regression.

During the last decade, due largely to the pioneering work of Leo A. Goodman of the University of Chicago, these methods have seen extensive use in sociology. But the application of log-linear methods is rapidly expanding to other areas of behavioral inquiry, even areas with strong experimental traditions. Unfortunately, at present, most didactic publications have been prepared largely for readers with recondite backgrounds in statistics and for the application of log-linear analysis to survey-oriented research. Unlike these publications, this book has been prepared for researchers and students who have modest backgrounds in statistics and who might very well have substantive interests in a field of inquiry with an experimental tradition.

To be more specific, this book is designed to be used by readers who have had only one *solid* course in applied statistical methods, which included the basic principles of factorial analysis of variance. To assist readers with modest preparation in quantitative studies, reviews of relevant topics in basic probability theory and traditional chi-square goodness-of-fit procedures are contained in Chapters 2 and 3. Students with strong backgrounds in statistics can skim over these preparatory chapters without prejudice.

Because log-linear analysis is relatively new and because its origins are quite diverse, several approaches can be taken to its formulation and application. The approach most familiar to this writer and hence the approach that is reflected here is that associated with Professor Goodman. Readers who have been exposed to Goodman's approach, however, will find the present treatment of the topic somewhat at variance with his writings and the writings of his associates. Aside from the modest requisites, experimental and ex post facto research are emphasized, not the survey. Expect, therefore, to see greater emphasis given to the synonymity between log-linear analysis and the analysis of variance, the "natural language" of psychologists and educators. For example, interpretation is more apt to be couched in terms of *effects* as opposed to *odds ratios*. Also, beginning in Chapter 5, attention will be directed almost exclusively to the study of group differences with respect to one or more dependent variables. Special contingency table models, known as logit models, will be used to estimate treatment effects and interaction effects as is done in the typical analysis of variance. In sum, many behavioral researchers will appreciate the frequent analogies drawn between log-linear analysis and the analysis of variance, and the marked emphasis afforded logit-model analysis, for such emphases are not only highly compatible with their past training but also lend themselves to the types of research typically conducted in their fields.

Numerous people helped in the preparation of the manuscript, and to these people I am deeply grateful. To Dr. Janet Rice of the Medical School at Tulane University go my thanks for her help with the topic of symmetric change discussed briefly in the last chapter. I am also most indebted to the many graduate students who participated in my seminar on log-linear analysis at The Ohio State University for their encouragement, ideas, and criticisms. And for their long hours and special help with the preparation of the manuscript, a special kind of thanks for Mrs. Beverly Doyon and Nancy J. Kadunc.

J. J. K.

Contents

List of Tables

List of Figures

ANALYZING
QUALITATIVE DATA

1

Qualitative Data
and Research

The analysis of qualitative/categorical data, data which are cross-tabulations within contingency tables, is an important component in both the study and conduct of behavioral research. Until recently, however, the statistical theory and computational techniques for the analysis of qualitative data were limited, for the most part, to data within the context of two-dimensional tables. For such tables, the use of chi-square goodness-of-fit procedures, developed early in the history of statistics by Karl Pearson (1900), has proved to be of immense value. Often, however, researchers possess information sufficient to structure contingency tables of more than two dimensions. Further, they often desire to subject these multiple dimensions to a comprehensive, simultaneous analysis. Unfortunately, prior to the 1970s the methodology that would permit the desired analysis was not refined sufficiently to the extent that it could be of assistance to the practicing researcher.

Then, through the pioneering efforts of Birch (1963), Bishop (1969), Fienberg (1970), Goodman (1970, 1971a, 1971b), Grizzle and Williams (1972), and Haberman (1972), among others, there emerged several systems of analysis designed specifically to accommodate the simultaneous analysis of data in multidimensional contingency tables. These systems of analysis are generically termed *log-linear contingency table analysis*. This book presents a didactic exposition of the logic and substantive application of log-linear analysis and does so within a framework that is similar to the analysis of variance (ANOVA). It is hoped that such an exposition will have the effect of increasing the number of practitioners who can both appreciate and use this relatively new statistical system in their research.

QUALITATIVE VARIABLES AND DATA

Ironically, terms most basic to an area of study often prove to be most difficult to define. Definitions are required, nevertheless, and because this volume deals with the study of qualitative variables and resultant qualitative data, some attempt must be made to clarify these terms. As a first pass, it could be said that a qualitative variable is a variable that yields qualitative data—but such a definition is somewhat tautological and uninformative in the absence of knowledge about qualitative data. Therefore, we attend first to the general nature of data, returning subsequently to a more meaningful discussion of qualitative variables.

The Nature of Data

In most instances, data gathered during the course of a behavioral study consist of *measurements*, numbers assigned to objects or events according to a set of rules. Implied by this definition of measurement is that there are a number of different rules used in the process of assigning numbers. Due largely to the scholarship of S. S. Stevens (1946), at present there is substantial agreement that the process of number assignment (i.e., measurement) is governed by four fundamental sets of rules. These four sets of rules give rise to four *scales of measurement*, or more simply, four types of data: nominal, ordinal, interval, and ratio data. Since these four types of data are discussed extensively in most introductory textbooks in statistics and measurement, only the principal features of each are reviewed below.

Nominal Data. Nominal scaling connotes classification. Briefly, the objects or events of concern are examined for underlying similarities (or differences) and subsequently grouped on the basis of observed *qualitative* distinctions. Examples of nominal scaling abound. For instance, nominal scaling is implemented when research subjects are classified according to their sex, sexual orientation, ethnic background, race, area of academic specialization, type of learning disability, etc. Now, if the numbers of subjects falling into respective categories of a variable such as sex or ethnic background are tallied, the resultant counts or frequencies constitute nominal data. Simply put, nominal data consist of frequencies observed within categories of a qualitative or nominal variable.

Notice that nominal scaling barely conforms to the definition of measurement presented earlier since in most instances the act of assigning numbers is optional; and when numbers are assigned to groups or categories, they serve merely as labels to facilitate coding or identification. Because qualitative categorization is so rudimentary, and since the act of assigning numbers is so arbitrary, many theorists do not seriously regard nominal scaling as a formal measurement process. Nominal scaling and nominal data, however, merit our serious consideration for not only is this manner of "measurement" frequently used in behavioral research, but also the analysis of this type of data constitutes the subject matter of this text. In sum, nominal data falls under the rubric of qualitative data.

Ordinal Data. A unidimensional *quantitative* continuum underlies ordinal scaling operations. Hence, unlike in nominal scaling, the assignment of numbers reflects degree rather than kind. With the requisite knowledge that some subjects (or units of interest) have more, or less, of a particular trait than others, subjects can be ordered from highest to lowest on the trait in question and numbers assigned consecutively to reflect this ordering. The assigned numbers, called *rank values*, are no longer just labels of convenience but instead denote relative position in an ordered array. Collectively, rank values are data, specifically ordinal data. It should be understood that the analytical techniques presented in this book are not generally applicable to ordinal data—at least not to ordinal data in their pristine form. There is, however, an exception: if subjects are first categorized largely on the basis of qualitative attributes (i.e., nominal scaling) and then the various categories are rank ordered on the basis of a quantitative dimension (i.e., ordinal scaling), the variable and associated frequencies within categories (i.e., data) may well be amenable to the techniques described herein. For example, suppose that on a questionnaire subjects were asked to indicate their approximate income on an ordered scale by checking one of several presented categories (e.g., less than $5,000, $5,000 to $9,999, $10,000 to $14,999, etc.); then the resultant data associated with this *ordered* variable can be treated by the methods we shall describe.

Interval Data and Ratio Data. A unidimensional *quantitative* continuum also underlies interval scaling and ratio scaling. In both cases, one has knowledge that some subjects possess more, or less, of a particular trait than others. But, unlike ordinal measurement, for these kinds of data one is able to specify how much more, or less, of the trait is possessed by various subjects.

Turning first to interval scaling, subjects can be placed on a continuum that is characterized by equal-interval numerical values. Consequently, meaningful differences between and among subjects can be inferred by simple subtraction of respective scale values. Examples of attributes that conform to interval scaling, and hence produce interval data, are: (a) achievement when based on classroom examination performance, provided that the examination contains a sufficient number of objective test items, (b) academic standing based on conventional grade-point averages, (c) typing proficiency as reflected by the number of errors made on a typing test, and (d) short-term memory as indicated by the number of trials required to reproduce a list of words.

Turning next to ratio scaling, in addition to units of equal size a natural or inherent "zero point" can be specified. Since the absence of the measured trait can be identified, a ratio comparison can be made between two measures. That is, it can be said that one value is so many *times* larger or smaller than another. Instances of ratio data are encountered frequently in everyday life: height, weight, and measures of volume are but a few examples. Unfortunately, this most mathematically tractable form of data is not encountered frequently within the fields of behavioral inquiry.

A widely recognized advantage of dealing with either ratio- or interval-appearing data is that they are often amenable to conventional *parametric* statistical analyses. For example, if one or more independent or predictor variables are being examined in relation to a dependent or criterion variable that can be assumed to be at least interval, then a parametric analysis will probably be appropriate. Depending, of course, upon the specific situation, the data analyst can select from among such techniques as the analysis of variance (ANOVA), analysis of covariance (ANCOVA), or bivariate or multiple regression. Should there be two or more interval response measures, the analyst will likely search out a multivariate technique such as linear discriminant analysis, multivariate analysis of variance (MANOVA), or canonical variate analysis. In either case, being able to treat response measure data as interval (or ratio) generally permits the researcher to use parametric procedures (as opposed to non-parametric procedures) and thus realize greater relative efficiency (i.e., increased probability of achieving statistical significance) and informational yield. Further, rarely is it ever justified to try to circumvent the use of parametric procedures by "scaling down" reliably gathered interval or ratio data so as to give it the appearance of ordinal or nominal data. But, as students of behavioral inquiry realize, often *all* variables of concern, both independent and dependent variables, are distinctively qualitative. Consequently, the parametric procedures cited above are either not appropriate or not as desirable as the procedures discussed in this book.

To this point we have established that the discussions to follow will be devoted exclusively to the statistical analysis of qualitative data, i.e., nominal data in either inherent or scaled-down form. Incidentally, a number of alternative terms are used to denote qualitative data. Common synonyms are attribute data, discrete data, categorical data, cross-tabular data, classification data, and enumeration data. And although it is often said that variety is the spice of life, variety can promote confusion in technical discourse. To minimize confusion, the term qualitative data will be used for nominal data throughout.

The Nature of Qualitative Variables

Qualitative variables give issue to qualitative data. And to repeat, we will be dealing only with qualitative variables. These variables may be *ordered* (e.g., grade level, income-level categories) or they may be *unordered* (e.g., ethnic background, religious affiliation); in most instances, we will be dealing with the latter type of variable. Distinction must also be made between a variable that is composed of only two categories and a variable that is defined by more than two categories. The former is commonly termed a *dichotomous* variable (or, sometimes, a binomial variable). The unordered variable Sex, consisting of a male and female category, is a dichotomy. However, a variable structured such that it subsumes more than two categories is called either a *polytomous* variable or a multinomial. Political Affiliation when defined by the categories of Republicans, Democrats, and Independents is an unordered polytomy. An example of an ordered polytomy is Class Standing defined by freshman→sophomore→junior→senior.

Looking ahead, the subject matter of this text consists of statistical methods, primarily methods associated with log-linear models, that are appropriate for use when research subjects are cross-classified on the basis of two or more qualitative variables and a simultaneous assessment of effects or relationships among these variables is desired. In addition to satisfying a number of conditions that we will discuss shortly, all variables entering into the analysis should be qualitative—ordered or unordered, dichotomies or polytomies.

BASIC CONSIDERATIONS

The intent of this section is twofold. The first is to enhance the reader's appreciation of the fact that the fundamental nature of a research problem (i.e., the types of research questions asked or the hypotheses advanced) carries with it important implications for how subjects should be sampled and how data obtained from subjects should be analyzed. The second is to review conditions that, within reason, should be satisfied prior to performing either the classical chi-square test developed by Pearson or the newer procedures associatied with the generation and evaluation of log-linear models.

To provide a context for the ensuing discussion of these points, consider, somewhat in isolation, the following two qualitative variables: Sex of Subjects, and Subjects' Attitude toward a Constitutional Amendment That Would Prohibit Abortions. Let us symbolically refer to the first variable (Sex) as the A variable. Further, let i be a subscript to denote the various categories subsumed by Variable A where i varies from 1 through a (i.e., $i = 1, 2, 3, \ldots, a$). Note that since A is a dichotomous variable here, $a = 2$. Our second variable (Attitude toward the Abortion Amendment) will be symbolized by B and subscripted by j, where $j = 1, 2, 3, \ldots, b$. Assume for the moment that B is also a dichotomy (hence $b = 2$). In sum, we have:

A: Sex of Subject	B: Attitude toward the Amendment
A_1 = females	B_1 = oppose
A_2 = males	B_2 = support

Now, in a typical research setting, one of two *modes of inquiry* will likely be adopted. That is, when viewed from afar, there are two general types of research questions that can be asked or, if justified, two general types of hypotheses that can be subjected to test: symmetrical and asymmetrical.

Symmetrical vs. Asymmetrical Inquiry

One can ask, for example, whether Variables A and B are independent—as opposed to being associated or correlated. Independence, or lack of association between A and B, implies that males and females respond in approximately the same manner when expressing their views on the abortion amendment; or, by the same token, subjects who have expressed either a positive or negative view are equally represented, in a proportional sense, by both males and females. Because a question (or hypothesis) dealing only with the presence or absence of an association between variables can be approached bilaterally, the mode of inquiry is *symmetrical*. Moreover, in the symmetrical mode, it follows that one would not designate one of the variables as the explanatory variable (or independent variable) and the other as the response variable (or dependent variable), for should an association be manifested between A and B, it may be that Variable A is the agent that accounts for the observed distribution in B, or B may be responsible for the observed distribution in A. Yet still, as often is the case, some variable or set of variables other than A or B may be influencing both A and B such that they vary systematically together. In any event, if the intent of the research is to determine whether there is an association between variables, or whether there is support for an a priori hypothesis that posits an association between two variables, then the mode of inquiry and subsequent analysis is said to be symmetrical.

In marked contrast is inquiry that seeks to determine whether subjects who fall into respective categories of one variable differ appreciably in their response to the other variable. For example, a researcher might ask whether the attitude of males and females is different with respect to the abortion-amendment issue. Here, Variable A assumes the status of an *explanatory* or independent variable while Variable B constitutes, in effect, a *response* or dependent variable. This perspective is distinctly undirectional or asymmetrical, not bilateral as was the case previously. Hence, the analysis to be applied can be termed *asymmetrical*. As we shall see, symmetrical inquiry and asymmetrical inquiry are related. If, for instance, Variables A and B are determined to be independent, it will then follow that there will be no material differences in the way that males and females respond to the abortion issue. Should A and B prove to be associated, however, the pattern of female response to the abortion question will be different from that observed for males.

In latter stages of the research process, it is likely that inferential statistical techniques will be used to pursue one of the two modes of inquiry just mentioned. Some inferential techniques—techniques such as simple correlation, multiple correlation, and canonical correlation—are used primarily to identify and test relations between or among variables. Still others—techniques such as ANOVA and MANOVA—serve to document statistically significant effects or differences between and among group means or, in the case of MANOVA, group centroids. Unfortunately, all too often a sufficient distinction is not made between these two modes of inquiry in discussions of qualitative data. We shall attempt to overcome this shortcoming, for, as has been mentioned, the adopted mode of inquiry carries with it major implications for sampling and data interpretation.

Random vs. Fixed Sampling

Recall that the function of inference in statistics is the generation of reasonable statements about *parameters* (population values) based upon careful examination and analysis of *statistics* (sample values). Recall further that requisite to the valid use of inferential statistics is both an adequate definition of the target population and access to a sample that is representative of this population. Moreover, random selection of subjects for inclusion in samples constitutes not only the ideal method to achieve representation but is also fundamental to the proper operation of the machinery of statistical inference.

But let us acknowledge that it is not always possible in practice to use random selection procedures to the extent that might be desired. When samples cannot be drawn at random, two options are available. The practitioner may elect to define the desired population to which inference will be made (the target population) and then proceed intelligently to use all relevant available information in an attempt to structure a sample that appears to be representative of the target population. (Structuring a judgment sample, however, is easier said than done.) A second option, and the only option available when the sample has been predetermined, is to study the sample at hand and then attempt to define a population from which the sample in question could have been drawn. Neither option is a satisfactory substitute for random sampling—yet research is an enterprise broader than the exact application of statistical methodology. And even if requisite conditions are not completely satisfied, statistical methods can still be of value. For example, with respect to the two nonrandom situations above, statistical inference may still be appropriate, but the producers and consumers of this research need to be aware that the statistical tests and estimates may be biased, due to biased sampling. Obviously, extreme caution should be exercised when advancing conclusions that go beyond the particulars of the immediate sample.

Returning to the theoretical level, if it is the researcher's intent to determine whether an association is present between two variables, complete use of randomization in sampling is called for. On the other hand, if the intent is to search for group differences, restricted use of randomization will suffice. With respect to the former (i.e., the symmetrical case), ideally one defines a population of interest and then proceeds to draw a simple random sample from the population. For our working example, assume that both the target and accessible populations consist solely of undergraduate students enrolled in a large university. If the purpose is to see whether sex and attitude toward abortion are related, obtaining a random sample by straightforward means constitutes the most appropriate plan. Not only will simple random sampling satisfy an important underlying condition associated with the statistical analysis of resultant data, but, as will be discussed, it will also yield *marginal* proportions for both Variables A and B which are *maximum-likelihood estimators* of respective proportions in the population. For the present, realize that the integrity of a symmetrical analysis rests heavily on the assumption that the sample is representative of the population and that random sampling is the best all-purpose method to achieve representation.

In contrast, if the intent is to determine whether differences (effects) exist between or among groupings of an explanatory variable (i.e., the asymmetrical case), stratified random sampling is generally preferable. Specifically, if we want to find out whether males and females hold different views on the abortion-amendment issue, then the respective number of males and females in the sample need not reflect the ratio of males to females in the background population. In fact, it is often desirable to *fix* in advance the number of males and females such that they are equally represented in the sample. Having equal, or approximately equal, numbers of subjects in categories of the explanatory variable has the salutary effect of promoting statistical efficiency, i.e., increasing the probability that statistical significance will be achieved. Again, this is not the best time to fully explain these notions. Suffice it to say that for an asymmetrical analysis, the researcher can use the categories associated with an explanatory variable (say, Sex) to stratify the population, and then draw samples of a desired size at random from within the strata. In sum, the respective frequencies of the explanatory variable can be *fixed* by the researcher, but the frequencies associated with the response variable should be free to vary within categories of the explanatory variable.

Cross-Classification of Subjects

Assume for our working example that: (a) the variables Sex and Attitude (for short) have been adequately defined so that, with respect to each, subjects can be unequivocally classified, (b) the essential nature of inquiry (symmetrical vs. asymmetrical) has been determined, (c) a sampling scheme appropriate to the nature of inquiry has been adopted (completely random vs. stratified random), and (d) information pertaining to Sex and Attitude has been obtained from each member of the sample. The next step is to marshal the sample information in preparation for descriptive and inferential analysis. The device used to organize and display qualitative data is a *contingency table.*

To construct a contingency table, the researcher first places subjects into the category belonging to Variable *A*, classifies them again on Variable *B*, and then crosses *A* and *B* in a manner similar to factorial design arrangements common to the analysis of variance. The result of this joint classification is a two-dimensional table. Technically speaking, when sample *frequencies* occupy the cells of the table, we have a contingency table, alternatively called a *table of cross-tabulations.*

Before we examine a contingency table, keep the following in mind. First, a sample member is associated with only one cell in the table; hence for now, repeated measurements, as are seen in certain ANOVA designs, are precluded. Second, frequencies or counts, not proportions or percents, are displayed in a contingency table. (It is often helpful, however, to convert observed frequencies into proportions which, when done, will give a *table of cell proportions.*) Finally, realize that contingency tables and corresponding tables of cell proportions are not limited to two dimensions. For example, if three qualitative variables are present, subjects can be cross-classified on all three variables, producing a three-dimensional contingency table. Eventually we will be working with three-, four-, and five-dimensional tables.

But first consider the type of table which is appropriate for our working example, a 2 X 2 contingency table, so labeled because the row variable (say Variable *A*) consists of two categories and the column variable (Variable *B*) also consists of two categories. Contingency tables of this type are sometimes called *fourfold tables,*

a descriptor that specifically informs us that the table contains four *elementary* cells. Fourfold tables have been of great interest to statisticians because, like classical experimental designs in which all factors consist of two levels, these structured configurations have a number of interesting properties and applications, some of which will be encountered below. A fourfold table containing hypothetical data obtained from a sample of 100 undergraduate students is presented as Table 1.1.

From a glance at Table 1.1, several inferences can be cautiously advanced. For example, if the 100 subjects ($n = 100$) in fact constitute a representative sample of the background population (i.e., all undergraduates in the university), and if the data are being examined from a symmetrical perspective, then it appears that there is an association between Sex and Attitude because males tend to be more favorably disposed to the amendment than are females. Approached from the other direction, subjects who do not support the amendment tend to be overly represented by members of the female group. From an asymmetrical perspective, where Sex is the explanatory variable, Table 1.1 data suggest a differential response by sex to the abortion issue. That is, males appear to favor the amendment in greater proportions than do females. Regardless of perspective, however, strong inferences concerning the state of affairs in the background population cannot be made at this point since we know that the apparent association, or male-female difference, could be the result of sampling error. To determine whether the data in Table 1.1 represent an outcome that is over and above that which could have occurred by chance, tests of statistical significance are needed.

TABLE 1.1. Cross Tabulations (Frequencies) by Sex and Attitude toward a Constitutional Amendment: An Example of a 2 × 2 Contingency Table

Sex of Student	Attitude toward the Amendment		Marginal Totals
	Opposed	Support	
Female	33	7	40
Male	37	23	60
Marginal Totals	70	30	100

Inferential Tests of Significance

Inference in statistical work is used to: (a) generate point estimates of and confidence intervals about population parameters (estimation), or (b) to assess the credibility of hypotheses pertaining to population parameters or conditions (hypothesis testing). By and large, the latter function—hypothesis testing—will receive the greatest emphasis in this book. Recall that three general steps are involved in the process of testing a hypothesis:

Step 1. A hypothesis is advanced in null form which either claims that a population parameter is equal to a specified value or maintains that a prescribed condition exists in the population of interest. In either case, it is the null hypothesis that is subjected to test.

Step 2. A sample (or samples) assumed to be representative of the population is drawn and an observed sample value or condition (statistic), corresponding to the value or condition offered under the null hypothesis, is assessed. In most instances, the comparison between that seen in the sample and that hypothesized by the null is accomplished within the context of a formula that yields a *test statistic* (e.g., an F statistic, a χ^2 statistic, etc.).

Step 3. The computed test statistic is related to the appropriate *sampling distribution*, a distribution of all possible test statistics that theoretically would result from repeated sampling under the null hypothesis. Almost always, sampling distributions are tabled (e.g., an F table, a χ^2 table, etc.). If the computed test statistic falls on an extreme tail of the sampling distribution (i.e., the region of rejection), we conclude that it is not likely that the sample in hand could have been drawn from the population posited under the null hypothesis. That is, grounds appear sufficient for rejection of the null in favor of acceptance of a logical alternative to the null. On the other hand, should the test statistic not fall in the region of rejection, evidence is insufficient for rejection of the null.

The steps overviewed above are used extensively in the analysis of qualitative data, and hence there will be ample opportunities to illustrate these familiar steps in the remaining chapters. However, the reader should be prepared to adopt a more flexible posture with respect to these inferential procedures as we approach discussions of log-linear analysis. As will be seen, a major feature of log-linear analysis is the fitting of statistical models, each model representing a different hypothesis, where the most appropriate model will be selected for subsequent interpretation and use. And, as the reader will discover, the process of model fitting and selection is not as prescribed as are classical decisions regarding the rejection of a null hypothesis. But then, therein lies the challenge and excitement of working with log-linear models.

A number of additional understandings merit brief mention in this introductory chapter. Most basic is knowledge of statistical procedures that are appropriate for various types of contingency tables. Consider, for example, a 2 X 2 contingency table such as that depicted by Table 1.1. For this type of table, if certain assumptions can be made and certain underlying conditions are met, the data analyst can select from among three analytical techniques: (a) a normal deviate test (z test) between two sample populations, (b) Pearson's classic chi-square test, or (c) a log-linear analysis. The choice of specific technique depends on the nature of inquiry (symmetrical or asymmetrical), and the amount and the nature of information desired. For a two-dimensional contingency table of size larger than 2 X 2, then either Pearson's chi-square or a log-linear analysis is appropriate. Finally, should the contingency table be of dimensionality greater than two, say a 2 X 2 X 2 table, then the log-linear approach is clearly preferable.

But irrespective of table dimensionality, or irrespective of whether a z test, a Pearsonian chi-square test, or log-linear analysis is being performed, associated with all of these techniques is a common family of sampling distributions. In a sense, all techniques cited belong to the same species in that the *multinomial law*, or an extension of this law known as the *product-multinomial law*, provides outcome probability distributions under null hypotheses. Technically, the multinomial law produces the appropriate *sampling distribution* for analyses that are symmetrical, while the product-multinomial law applies to the asymmetrical case. The next chapter contains a discussion of these laws.

Having established, at least initially, that *exact* sampling distributions of observed sample frequencies are provided by either the multinomial or product-multinomial law, it must be pointed out that in most of our work we will not be directly consulting these exact sampling distributions. We will work instead with sampling distributions resulting from the distribution of chi-square statistics, distributions that *approximate* the exact sampling distributions given by the multinomial and product-multinomial law. In other words, in the presence of reasonably large samples, the computed test statistic will be a goodness-of-fit chi-square statistic which in turn will be related to a chi-square sampling distribution, the latter serving as an approximation to the exact multinomial or product multinomial. Again, considerable explication of these relations is needed, and this will be provided in the next chapter.

Finally, although much has been made of the distinction between the symmetrical and asymmetrical cases, and the differential consequences thereof, when it comes to the analysis both approaches lead to the same estimates of population cell frequencies and the same goodness-of-fit test statistics. This distinction, therefore, has consequences for sampling and the interpretation of results but not for the mechanics of the analysis.

CONCLUDING REMARKS

It is hoped that through this chapter the reader now possesses a clearer understanding of the nature of qualitative data and qualitative variables and, as a result, possesses a clearer idea as to the type of research that will be addressed in this book. Moreover, the reader should now appreciate better the differences between research that seeks to explore relations among qualitative variables and efforts that attempt to investigate differences among groupings of subjects with respect to their responses to a qualitative variable, since these distinct modes of inquiry affect both the manner in which samples are obtained and, subsequent to a common goodness-of-fit analysis, the manner of conclusions to be drawn. In addition, it has been pointed out that when data emanate from contingency tables of three or more dimensions, there are serious limitations associated with traditional approaches to the analysis of contingency-table data.

Lastly, readers who presently have a good grasp of multinomial probability theory and of classical chi-square goodness-of-fit procedures may elect to proceed immediately to Chapter 4 where they will be introduced to log-linear analysis. Prior to advancing to Chapter 4, however, readers who currently possess some understanding of the log-linear approach may find the discussion contained in the final section of Chapter 7 of interest. The discussion in that section centers about the relative merits of log-linear methods, particularly with respect to alternative procedures. For those readers who are not yet ready for either this discussion or for an introduction to log-linear methods and could profit from a review of traditional chi-square and related procedures, the next two chapters are provided.

2

Statistical Foundations

Populations, samples, and sampling (probability) distributions—each of these distributions plays a special and important role in statistical inference. Enhanced knowledge of *populations*, of course, constitutes our prime and ultimate objective. And recall that it is the intent of inference in statistics to either (a) generate estimates of unknown population values, called *parameters*, and, if desired, to place confidence intervals about these estimates, or (b) subject to test statements concerning the value or equivalence of parameters. Generating point estimates (of parameters) and determining the accuracy of those estimates is known as estimation. The latter type of inference— inference in which a null hypothesis is subjected to test—is called, simply, hypothesis testing.

In either case, parameter estimates or decisions about the viability of null hypotheses are based on a careful study of a *sample* that is assumed to be representative of the population of interest. From the distribution of sample observations, *test statistics* are computed. (The computation of a t statistic or a chi-square statistic from sample data constitute but two of many specific examples that could be offered.)

Last, but far from being least, is a distribution that provides the crucial linkage between populations and samples and thus enables us to posit either estimates or conditional conclusions about the status of null hypotheses, namely a mathematical probability distribution often called a *sampling distribution*.

Sampling distributions are linked to specific sample statistics. The sampling distribution of a particular statistic, such as a *t* or chi-square, may be viewed as a *relative frequency distribution* made up of an infinitely large number of statistics independently computed from samples (of size *n*) which have been drawn from the population (or populations) specified by the null hypothesis. In short, sampling distributions are relative frequency distributions of all possible test statistics resulting from repeated sampling under the null. As a consequence, sampling distributions are used to assess the likelihood that a test statistic, computed on a particular sample, could have emanated from the population specified by the null hypothesis. If it happens that a test statistic fits snugly in the middle of its sampling distribution, then there is no basis for rejecting the null hypothesis. On the other hand, if the sample test statistic falls well into the tail of the appropriate sampling distribution, then available evidence suggests that it is *not* likely that the sample in question was drawn from the null population. And to the extent to which the computed test statistic represents an unlikely occurrence under the null, grounds are sufficient to reject, at least conditionally, the null hypothesis.

This chapter presents an overview of the sampling distributions that describe qualitative response. Discussed first is the distribution of dichotomies (binomial variables) and the *binomial law* which specifies the nature of sampling distributions associated with these variables. Extensions of the binomial law, namely the *multinomial* and *product-multinomial laws*, will be considered subsequently, for these extensions describe the sampling distributions of polytomous variables. Finally, this review chapter closes with a brief discussion of *maximum-likelihood estimation*—the type of estimation that is appropriate when response is qualitative, and the method of estimation most often used today to estimate parameters in log-linear models.

The topics of this chapter may appear at first to be far removed from direct, practical application since the emphasis here is on basic theory, not "nuts-and bolts" application. In fact, readers most anxious to begin applying log-linear methods in their work can skim through this chapter without incurring deficiencies that would make the reading of subsequent chapters more difficult. But at some point in the process of learning about and applying log-linear methods, even the most application-oriented researcher will want to become more comfortable with the mathematical-statistical underpinnings of qualitative data analysis. Aside from satisfying the intellect, obtaining at least an elementary appreciation for the statistical bases of qualitative analysis can only lead to application with greater confidence, authority, and possibly creativity. In other words, at some point the reader is encouraged to tackle the contents of the following sections so that the nagging, underlying mysteries often associated with the analysis of qualitative data can be dispelled.

DICHOTOMIES AND THE BINOMIAL LAW

The probability theory associated with dichotomies was established in the early part of the eighteenth century largely in an attempt to predict the outcomes of games of chance. James Bernoulli (1654–1705), in particular, is credited with the development of the majority of basic principles and mathematical proofs pertaining to dichotomous events. Before we attempt to apply Bernoulli's work to the dichotomous outcomes of behavioral studies, a review of several basic principles may prove helpful.

Bernoulli Trials and Resultant Sequences of Events

A Bernoulli *trial* yields an outcome, or event, that is unequivocally dichotomous. The toss of a standard coin which can only result in a "head" or a "tail" constitutes such an event. The roll of a six-sided die whereby a successful roll is considered to be the appearance of a six—all other outcomes are deemed a failure—is a Bernoulli trial. Inquiring of a research subject whether he or she is either in favor of, or opposed to, a constitutional amendment that would prohibit abortion is also a Bernoulli trial if the response is structured so that it is dichotomous.

In both gambling and research, however, our interest usually extends beyond the outcome of a single trial. Indeed, multiple trials leading to multiple outcomes generally are of interest. To accommodate this, therefore, let *n* stand for the number of trials of interest and, unless otherwise stated, assume that the *n* ensuing events are not only *mutually exclusive* (in a single trial, only one specified event can occur) but also *independent* (the occurrence of a specified event from one trial is not influenced by the occurrence of any other). We will have occasion later to discuss exclusivity and independence at greater length.

For now, we seek answers to the following three questions: (a) How many distinct sequences (permutations) are possible within *n* independent Bernoulli trials? (b) What is the probability of observing a sequence that contains a specified (desired) number of outcomes?

(c) How many distinct sequences contain a specified number of desired outcomes? It is hoped that by systematically answering these questions the foundations for understanding the binomial law will have been laid.

We begin with an example based on the roll of a fair six-sided die. Assume that the desired outcome is the roll of a six; the five remaining outcomes are not desired. Now, the probability of observing a six in one roll of a die is .1667 (rounded to the fourth decimal place), and under normal circumstances this probability will not change over successive rolls. This establishes a most important a priori condition for our work—namely, prior knowledge of the probability of the desired outcome for a single trial and hence for successive trials. Further, for Bernoulli trials knowledge of the probability of the desired outcome determines the probability of the undesired outcome. Therefore, if we let the numeral 1 denote the desired outcome and let the numeral 0 denote an undesired outcome, the respective probabilities for a single trial or roll of a die are:

P_1 = probability of observing a 1 = .1667

P_2 = probability of observing a 0 = .8333

But our concern is with multiple trials. Consider, therefore, four trials ($n = 4$) and a possible sequence or ordering of outcomes whereby a six results from the first roll of the die, a non-six results from the second roll, then another non-six appears, and the last roll produces a six. In our notation, the *distinct sequence* (permutation) described is

Seq: 1, 0, 0, 1

This sequence or permutation is, of course, but one of a number of distinct sequences that could result from four rolls of the die, which brings us to the first of our three questions: How many distinct sequences are theoretically possible from executing n independent trials?

To answer this question, let k stand for the number of mutually exclusive outcomes that are possible within the context of a single trial. In general, k is an integer greater than 1; but for a Bernoulli trial, k by definition is equal to 2. Now, from previous course work recall that the number of distinct sequences or permutations in n independent trials can be determined by

No. of Seq $= k^n$ $\qquad$ (2.1)

For Bernoulli trials, Eq. (2.1) may be written

No. of Seq $= 2^n$ $\qquad$ (2.2)

and since 2 raised to a power of 4 equals 16, there are 16 distinct sequences that could come about in four rolls of the die. (These 16 distinct sequences are given in the second column of Table 2.1.)

The second of our three questions involves determining the probability of a distinct sequence being observed. To calculate this value, prior knowledge of the probability of a desired outcome in a single trial must be known. For our example, this is known—i.e., $P_1 = .1667$. Calculations will be facilitated if we let the symbol f_1 stand for the number of desired outcomes (i.e., sixes) in the sequence and let f_2 represent the number of undesired outcomes (i.e., non-sixes) in the sequence. Thus, for the sequence in question (No. 8 in Table

TABLE 2.1. Distinct Sequences and Their Probability of Occurrence for Four Bernoulli Trials

Sequence No.	Sequence	Probability
1	1 1 1 1	.0008
2	1 1 1 0	.0039
3	1 1 0 1	.0039
4	1 0 1 1	.0039
5	0 1 1 1	.0039
6	1 1 0 0	.0193
7	1 0 1 0	.0193
8	1 0 0 1	.0193
9	0 1 1 0	.0193
10	0 1 0 1	.0193
11	0 0 1 1	.0193
12	1 0 0 0	.0965
13	0 1 0 0	.0965
14	0 0 1 0	.0965
15	0 0 0 1	.0965
16	0 0 0 0	.4822

Note. The a priori probability of observing a 1 has been established as .1667.

2.1), $f_1 = 2$, and $f_2 = n - f_1 = 2$. Now, for Bernoulli trials in general, the probability of observing a distinct sequence is given by

$$P(\text{Seq}: \dots) = (P_1)^{f_1} (P_2)^{f_2} \qquad (2.3)$$

Substituting into Eq. (2.3) the parameters of our working illustration and performing the indicated algebra gives

$$P(\text{Seq}: 1, 0, 0, 1) = (.1667)^2 (.8333)^2$$
$$= (.0278)(.6944)$$
$$= .0193$$

Thus, if we were to roll four dice a large number of times, in the long run we would expect that the prescribed sequence would result about 2% of the time. This computed probability value is presented in Table 2.1 along with the outcome probabilities of the remaining 15 distinct sequences.

Examination of Eq. (2.3) and Table 2.1 reveals that not only does the distinct sequence in question have an outcome probability of .0193, but all sequences in which $f_1 = 2$ and $f_2 = 2$ also share this outcome probability value. Therefore, for a specified number of desired outcomes (i.e., for a given value of n), Eq. (2.3) provides the outcome probability irrespective of the temporal or linear arrangement of the sequence. Or, more succinctly, Eq. (2.3) gives the probability of any *combination* of n dichotomous outcomes containing f_1 outcomes, a fact that the reader is encouraged to verify.[1]

We will now attempt to answer the third of our three questions, the question concerning the number of possible sequences or combinations that will manifest a specified number of desired results. The number of ways (combinations) of choosing f_1 things (observing f_1 outcomes) in a series of n Bernoulli trials can be found by

$$\binom{n}{f_1} = \frac{n!}{(f_1!)(f_2!)} \qquad (2.4)$$

Let us use this equation to determine how many sequences or combinations of four Bernoulli trials ($n = 4$) will contain two sixes

---

[1] In working with Eq. (2.3), realize that a number raised to the power of zero equals unity.

($f_1 = 2$). Remembering that 4! (read, four factorial) is $4 \times 3 \times 2 \times 1$, substituting into Eq. (2.4), and solving gives

$$\binom{4}{2} = \frac{4 \cdot 3 \cdot 2 \cdot 1}{(2 \cdot 1)(2 \cdot 1)} = \frac{4 \cdot 3}{2 \cdot 1} = \frac{12}{2} = 6$$

Incidentally, the above result is supported by the information shown in Table 2.1. As an exercise, it is suggested that Eq. (2.4) be used to verify that

$$\binom{4}{1} = \binom{4}{3} = 4$$

and that

$$\binom{4}{0} = \binom{4}{4} = 1$$

The Binomial Law

To this point, it has been established that if we know the probability of a desired outcome for a Bernoulli trial (e.g., $P_1 = .1667$), then for a series of independent trials (i.e., $n > 1$):

1. Eq. (2.2) will give the number of permutations or distinct sequences.
2. Eq. (2.3) will generate the mathematical probability that a sequence will occur which contains f_1 desired outcomes.
3. Eq. (2.4) will reveal the number of sequences that contain f_1 desired outcomes.

We should now be able to use the formulations above to write an equation to give the probability of obtaining a given outcome irrespective of distinct sequence.

Recall that we have established that 2^n distinct outcome sequences are possible. Moreover, some of these sequences will contain precisely f_1 desired outcomes. (Only if $f_1 = n$ or if $f_2 = 0$ will there be only one distinct sequence containing precisely f_1 desired outcomes.) Now, if we were to select just one of these desired-outcome sequences and then determine its probability of occurrence, it would be $P_1^{f_1} P_2^{f_2}$. But, with the exceptions noted above, there will be more than one sequence or combination of events containing f_1 outcomes. Hence, if the probability of observing a single desired sequence is $P_1^{f_1}P_2^{f_2}$, but there are

$$\binom{n}{f_1}$$

such sequences to be observed, it follows that the probability of observing a sequence containing f_1 is

$$(P_1^{f_1}P_2^{f_2})\binom{n}{f_1}$$

More formally, the probability of observing f_1 desired outcomes, given n trials, is

$$P(f_1, f_2 | n) = \frac{n!}{(f_1!)(f_2!)} P_1^{f_1}P_2^{f_2} \tag{2.5}$$

Equation (2.5) represents the sought-after binomial law, the law that governs the sampling distributions of independent dichotomous responses.

Let us use the binomial law to calculate the probability of observing two sixes in four rolls of a fair die. We substitute the appropriate values into Eq. (2.5) and solve:

$$P(2, 2 | 4) = \frac{4!}{(2!)(2!)} (.1667)^2(.8333)^2$$

$$= (6)(.0193) = .1158$$

Probabilities for the remaining combinations, calculated by similar means, are summarized below:

$$P(0, 4 | 4) = (1)(.4822) = .4822$$

$$P(1, 3 | 4) = (4)(.0965) = .3860$$

$$P(2, 2 | 4) = (6)(.0193) = .1158$$

$$P(3, 1 | 4) = (4)(.0039) = .0156$$

$$P(4, 0 | 4) = (1)(.0008) = .0008$$

Now that we know the probabilities associated with all possible outcomes (all values of f_1) for the working example, it is possible to construct a relative frequency distribution of respective outcome probabilities. A histogram, depicted in Figure 2.1, is most appropriate here because the distribution is discrete, not continuous. Of course, the distribution depicted is a sampling distribution, the sampling distribution for any dichotomous response variable with parameters $P_1 = .1667$ and $n = 4$. Notice that this particular sampling distribution is not symmetric about .1667, the *expected value* or mean of the distribution. Only if P_1 is set at .5000 will the distribution of binomial events turn out to be symmetric about .5000, its expected value.

Concrete Illustrations

To apply our work so far, imagine that we suspect that at a crucial moment in a game of chance the croupier has surreptitiously substituted a "loaded" die for a fair die. Specifically, we suspect that at critical points involving four consecutive rolls of a single die, the

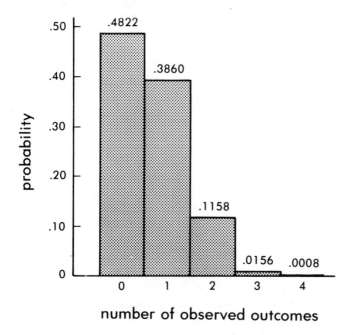

Figure 2.1. Expected relative frequency of observing a six in four rolls of a fair die.

croupier's chances of winning the game are being enhanced by a die that in the long run will produce a disordinate number of sixes. Assume that we have but one opportunity, one critical series of four rolls, to test our hypothesis that achieving a six on the roll of this die is greater than .1667. (Note that this is analogous to being able to obtain only a sample of size four in a behavioral study.) Further, since absolute certainty is beyond our grasp, we will be willing to assume a 5% risk of concluding that the die is loaded when in fact it might not be. (In short, we establish the level of significance at .05). To conduct the test, the following steps are executed:

1. We state the null hypothesis, the hypothesis to be subjected directly to test:

$H_o:P_1 = .1667$

2. We establish a reasonable level of significance, say $\alpha = .05$.

3. We draw from the population of critical throws a representative sample of events. For the case at hand, this constitutes the observance of four critical rolls of the suspected biased die (i.e., $n = 4$).

4. We calculate the appropriate sample statistic or statistics. Here, assume that the four critical rolls produced a six, a two, a six, and then another six; using our original notation,

Seq: 1, 0, 1, 1

and thus $f_1 = 3$.

5. We relate the observed sample statistic to an appropriate sampling distribution to determine the probability associated with the observed sample outcome. That is, the observed outcome ($f_1 = 3$) will be compared to the probability distribution displayed in Figure 2.1.

6. We posit a conditional conclusion relative to the tenability of the stated null. Here, the probability of observing *three or more* sixes in four rolls of a fair die is .0156 + .008 = .0164, an unlikely, albeit possible, result. But since earlier we decided that any sample outcome that could be expected less than 5% of the time would lead us to reject the null, we do reject in favor of a logical alternative. We conclude, therefore, that $P_1 > .1667$. The die was loaded during the four critical rolls.

Consider a more realistic example, one in which a civil service examination was given to police officers for the purpose of selecting candidates eligible for promotion. In this example, exam results for individuals were scored dichotomously: they either passed or failed the examination. Of those who took the exam, 30% were members of a minority group. Only 1 minority group member was present on the list of 10 who passed the exam, however, which prompted the concern that the examining procedure had an "adverse impact" on minority members of the department. It could be argued, for example, that if a population of officers who took the exam consisted of 30% minorities, and if there is absolutely no basis for even suspecting adverse impact, then in the long run 30% of those who pass the exam would be members of the minority. But the present sample of size 10 contained only 1 successful minority member. Let us, therefore, undertake to assess the likelihood of observing only a 10% minority success rate assuming that the examination does not impact adversely on minorities.

The analysis proceeds as follows:

1. A statement of the null. In the population represented by the present sample, the probability of minority success is at least .30; that is, $H_o : P_1 \geqslant .30$.

2. An a priori determination of alpha. Set $\alpha = .05$.

3. An examination of relevant sample statistics. Here, when $n = 10$, the outcome of interest—minority success—was $f_1 = 1$.

4. A comparison of the observed sample result (i.e., $f_1 = 1$) with results to be expected given that the null hypothesis is true. A sampling distribution describes expected results under a null. To generate the sampling distribution that is relevant to the working example, Eq. (2.5) is used. Specifically, Eq. (2.5) can be used repeatedly to generate at least the lower tail of the exact discrete sampling distribution for $n = 10$ and dichotomous outcomes. For example, the most extreme lower-tail outcome is one whereby no members of the minority and 10 of the majority pass the exam. The probability of this extreme outcome is

$$P(0, 10 | 10) = \frac{10!}{(0!)(10!)} (.30)^0 (.70)^{10}$$

$$= (1)(1)(.0282)$$

$$= .0282$$

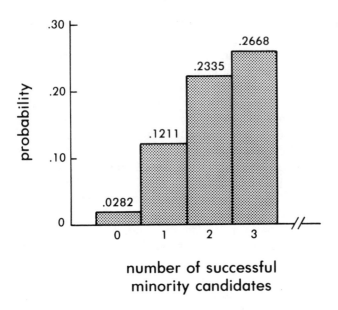

Figure 2.2. Lower-tail probabilities for a binomial sampling distribution where $n = 10$ and $P_1 = .30$.

Admittedly, on an examination where $P_1 = .30$, it is possible to have no successful minority candidates on a list of 10, but in the long run this outcome will occur less than 3% of the time. Equation (2.5) is now used again to determine the next most extreme outcome, i.e., $P(1, 9|10)$, and subsequently all other lower-tail outcome probabilities needed to perform the test. The results of this work, which the student should at least partially verify, are summarized in Figure 2.2.

5. A conditional conclusion is advanced with respect to the viability of the null hypothesis. In the present case, a comparison of the observed sample outcome ($f_1 = 1$) with the sampling distribution of Figure 2.2 indicates that the probability of observing 1, or fewer, minorities in 10 is $.0282 + .1211 = .1493$. But since the probability of observing 1 or fewer minorities is greater than .05—the level of significance established for this research—present evidence is not sufficient to lead to the rejection of the null.

Large-Sample Approximations to the Binomial

Consider again the results of a promotional examination but instead of 10 officers on the list of successful candidates, assume that the list contained 30 candidates ($n = 30$). Assume further that the same percentage of minority group members was represented on the list; that is, 3 of the successful candidates, or 10%, were minorities. An exact test of the null hypothesis that $P_1 = .30$ would proceed as before, but greater computational effort will be required to generate the lower tail of the sampling distribution. Specifically, it will be necessary to employ Eq. (2.5) to calculate $P(0, 30|30)$, then $P(1, 29|30)$, then $P(2, 28|30)$—and such additional probabilities as will be required to generate at least the region of rejection of the sampling distribution. In fact, six probability calculations are needed to construct the 5% lower-tail region of rejection for this problem. All things being equal, the larger the sample, the greater the labor required to generate the region of rejection. For example, for a sample of size 300 the computational labor required is, for all practical purposes, prohibitive.

Fortunately, if sample sizes are sufficiently large the use of the binomial theorem to generate sampling distributions can be circumvented through use of the *central limit theorem*. This theorem holds that as n becomes infinitely large, the normal distribution becomes the limiting form of a binomial distribution for a fixed P_1. Therefore, as is shown in many basic textbooks in statistics, the standard normal probability curve serves as an excellent approximation to the discrete binomial distribution when samples are sufficiently large.

A question that is often asked is how large a sample must be to justify the use of the normal distribution in place of the binomial. In response, a specific sample size cannot be given because the adequacy of the normal as a substitute for the binomial depends on the relationship between the parameter contained in the null (i.e., P_1) and the size of the sample (n). Briefly, as the null parameter approaches .50 (hence P_2 approaches .50), a sample size as small as 10 will produce a discrete binomial distribution that, for practical work, is reasonably approximated by the continuous normal distribution. But as the null parameter deviates from .50, a larger n is required for reasonable approximation by the normal. For example, with respect to the working illustration in which $P_1 = .30$ and $n = 10$, the symmetric unit normal is not an acceptable substitute for the exact asymmetrical binomial. (If sketched, a visual comparison of these two distributions would reveal unacceptable discrepancies particularly in the tails of the distribution.) For our working illustration where $P_1 = .30$ but $n = 30$, however, the normal does sufficiently approximate the binomial for most applications.

As a practical guide, the *rule of five* is offered as a means to judge whether a normal distribution, and hence a z test, can be used in place of the more cumbrous binomial procedures. This loosely defined "rule" holds that the normal distribution is a reasonable approximation to the binomial when the product of the null parameter and sample size exceeds 5. Applying the rule to our initial example ($n = 10$) produces a product whose value is

$$nP_1 = (10)(.30) = 3.0$$

and, since 3 is less than 5, the use of a "large-sample approximation" is discouraged. In contrast, when applied to our second example ($n = 30$),

$$nP_1 = (30)(.30) = 9.0$$

an outcome that justifies the use of an approximate z test.

Since the unit normal can be used to approximate the binomial for the working example involving 30 successful testees, let us pause for the purpose of performing a z test on the null $P_1 = .30$, a test that will require far less computational labor than the exact binomial test. The major steps are:

1. Statement of the null, $H_o : P_1 \geqslant .30$.
2. Establishment of the level of significance, say $\alpha = .05$ for a lower-tail test.
3. Calculation of relevant statistics on the representative sample. On the list of successful candidates ($n = 30$), 3 were minorities; therefore, $f_1 = 3$. Thus, the corresponding observed sample proportion, denoted by p_1, is

$$p_1 = \frac{n}{f_1} = \frac{3}{30} = .10$$

4. Computation of the test statistic, an approximate z statistic in this instance. In many basic texts, the standardized form of f_1 (or np_1) for any size sample is essentially given by

$$z = \frac{f_1 - nP_1}{\sqrt{nP_1P_2}} = \frac{f_1 - nP_1}{(nP_1P_2)^{1/2}} = \frac{p_1 - P_1}{(P_1P_2/n)^{1/2}} \tag{2.6}$$

Be sure to notice that the radical sign indicating the degree of the required root (the square root in this case) has been replaced by the equivalent operation of raising quantities enclosed within parentheses to corresponding fractional exponents. Henceforth, extensive use of this versatile and economical operation will be made in this book. Returning to the test, substitution of numerical values associated with the working example into the right-most version of Eq. (2.6) results in

$$z = \frac{.10 - .30}{[(.30)(.70)/30]^{1/2}} = \frac{-.20}{.0837} = -2.39$$

5. Comparison of the test statistic computed above to the *critical z* value which defines the lower-tail region of rejection on the unit normal sampling distribution $(z_\alpha = -1.65)$. In this case, the comparison indicates that a significant result has been achieved.

6. The conclusion: reject the null in favor of the alternative hypothesis that the proportion of minorities passing the examination is less than .30. Evidence exists in support of the suspicion that the examination has had an "adverse impact" upon minorities.

Before considering extensions of the binomial theorem, three brief but important points should be made. The first is that there is an intimate link between the unit normal probability distribution and a chi-square (χ^2) probability distribution when the latter is distributed on a single degree of freedom (i.e., $\nu = 1$). Specifically, $z^2 = \chi^2$ when the latter has but one degree of freedom. Thus, for dichotomous responses in the presence of large samples, the chi-square distribution defined by one degree of freedom also serves as an approximation to the binomial. Think, therefore, of the unit normal distribution as a special case of the more generic chi-square, a chi-square distributed on a single degree of freedom.

The second point to be made is that since binomial distributions are discrete (and hence "step-like"), and their normal or chi-square approximations are continuous (and hence smooth), the correspondence between these two types of distributions at certain locations on these curves becomes less precise as sample sizes become small. Such imprecisions can be partially ameliorated by affecting an adjustment when using either the normal deviate or chi-square test. That is, when the rule of five gives a product that is greater than 5 but less than 10, greater accuracy can be achieved if a *correction for continuity* is made. For both a discussion of the continuity problem and a presentation of computational procedures, consult Hays (1973, pp. 308–309).

Finally, neither the unit normal nor the single-degree-of-freedom chi-square are good approximations to the binomial when the P_1

parameter (or, conversely, the P_2 parameter) is so small that the rule of five cannot be satisfied even in the presence of a large sample. A case in point would be a test on the null hypothesis that $P_1 = .05$ using a large sample of say 60. For this and similar cases, there exists another probability distribution—the Poisson distribution—which is a better approximation of the binomial distribution.

The first two points cited above will be developed in greater detail in the next chapter, and the continuity problem is touched upon in Chapter 7.

POLYTOMIES AND THE MULTINOMIAL LAW

The rationale and the formula for binomial distributions can be generalized to trials that have more than two outcomes. For example, suppose that instead of classifying subjects who passed a promotional examination in a dichotomous manner (i.e., minority or majority), subjects were classified in a polytomous manner. To be more specific, assume that subjects were classified as being Black, Caucasian, or Hispanic. The generalization that describes the distribution of a polytomous outcome such as this is known as the multinomial rule, law, or theorem.

The Multinomial Rule for a Polytomous Variable

If we again let k stand for the number of outcomes per trial (hence, $k = 3$ in the example above), then the multinomial equation can be written as

$$P(f_1, \ldots, f_i, \ldots, f_k | n) = \frac{n!}{\displaystyle\prod_{i=1}^{k} f_i!} \prod_{i=1}^{k} P_i^{f_i} \qquad (2.7)$$

Equation (2.7) is in fact a straightforward extension of Eq. (2.5). The notation

$$\prod_i^k$$

denotes the *product operator*, which functions much like the summation operator Σ except that constituent factors (frequencies or proportions) are multiplied, in this case beginning with subscript $i = 1$ and terminating with $i = k$.

Let us exercise this equation for a situation in which the population of test-takers is 70% Caucasian, 20% Black, and 10% Hispanic. Assume that 10 test-takers passed the exam ($n = 10$) of which 8 were Caucasian and 2 were Black. Given the population parameters cited above, what is the probability that a list of 10 passing candidates will contain 8 Caucasians, 2 Blacks, and no Hispanics, assuming as we have done before that the examination does not impact adversely on members of these respective groups? Substituting into Eq. (2.7) the numerical values peculiar to this example, and remembering that 0! and $(.10)^0$ are both equal to unity, we obtain

$$P(8, 2, 0 | 10) = \frac{10!}{(8!)(2!)(0!)} (.70)^8 (.20)^2 (.10)^0$$

$$= (45)(.0576)(.0400)(1)$$

$$= .1038$$

Since the probability of the observed outcome is in excess of .10, and since by traditional criteria this probability itself is not small enough to justify rejection of the null hypothesis

$$H_0 : P_1 = .70, P_2 = .20 \qquad (\text{hence}, P_3 = .10)$$

attempts to generate additional lower-tail probabilities for outcomes more indicative of adverse impact are not needed.

As we have just seen, it is relatively easy to calculate the probability of a *specific* polytomous outcome. This cannot be said, however, for the generation and subsequent graphing of multiple outcomes associated with a sampling distribution. Just generating the relevant tails of the sampling distribution, for determining the respective probabilities of sample outcomes more deviant from a null than the observed outcome, can be a most arduous task, especially for higher values of n and k. Fortunately, there is a more convenient sampling distribution that can be used in place of the exact multinomial whenever samples are reasonably large. For large samples, the chi-square distribution provides a good approximation to the multinomial, as will be discussed at greater length in the next chapter. We simply point out for now that had our sample been larger, say $n = 50$, a chi-square distribution could have been used in place of the exact multinomial to perform an initial test on the null hypothesis.

The Multinomial Law and Contingency Tables

Single qualitative variables have been considered up to this point. However, for the most part, our interest and work will be directed toward situations in which two or more qualitative variables will be crossed to form a design structure, a contingency table such as was seen in Chapter 1. To facilitate discussion of contingency tables, we will first clarify our system of notation.

As in the previous chapter, let A be the generic symbol for the row variable. Further, we denote the constituent categories or levels of Variable A by the subscript i, where $i = 1, 2, \ldots, a$. Let B stand for the second variable, the column variable; this is subscripted by j, where $j = 1, 2, \ldots, b$. Generally, the uppercase letters P and F will be used to denote population proportions and frequencies, respectively. Thus, P_{ij} represents a parameter, specifically the proportion associated with the ith level of A and jth level of B. Corresponding sample proportions and frequencies will be indicated by lowercase letters; hence, p_{ij} and f_{ij} stand respectively for the observed proportion and frequency in the cell located in the ith row and jth column of the table. Additional notation will be introduced as needed.

When qualitative variables are crossed and contingency tables are formed, exact sampling distributions are usually described by either the multinomial law or a variation of this law known as the product-multinomial law. The sampling model determines, in large measure, the specific choice of sampling distribution. Two sampling models, random and fixed, were mentioned in Chapter 1. Briefly, the multinomial rule governs the exact distribution of sample statistics when *simple* random sampling is appropriate, while the product-multinomial law is the basis for sampling distributions when *stratified* random sampling is employed. These points merit greater elaboration.

If a sample of size n is drawn at random from a population and sample members are cross-classified on the basis of two (or more) qualitative variables, then, as noted, the straightforward version of the multinomial law is the basis for the sampling distribution. Said differently, if all we do is to arbitrarily fix (establish) the overall size of the sample (n), thus permitting selected subjects the opportunity to be placed in any one of the table's ab cells—depending, of course, on how they are jointly classified—then Eq. (2.7) can be used either to (a) determine the outcome probability of a given set of cross-classified data or (b) generate a complete sampling distribution.

Of course, to use Eq. (2.7) to perform the work indicated above, either the elementary cell parameters must be known or they must be assumed under a null hypothesis. For example, using the data of

Table 1.1, if we desire to test the null hypothesis that the variables Sex and Attitude toward the Abortion Amendment are *independent* (not correlated), we shall learn from the next chapter that the population proportions that reflect this independent state for these data are: $P_{11} = .28$, $P_{12} = .12$, $P_{21} = .42$, and $P_{22} = .18$. Referring back to Table 1.1, we are reminded that the four elementary cell frequencies observed for the sample of size 100 were: $f_{11} = 33$ ($p_{11} = .33$), $f_{12} = 7$ ($p_{12} = .07$), $f_{21} = 37$ ($p_{21} = .37$), and $f_{22} = 23$ ($p_{22} = .23$). Now, to determine the exact probability of achieving these observed elementary cell frequencies, given that the two variables are independent, Eq. (2.7) can be formidably implemented as suggested below:

$$P(33, 7, 37, 23 \mid 100) = \frac{100!}{33! \cdot \ldots \cdot 23!} (28^{33} \cdot \ldots \cdot 18^{23})$$

However, the solution to the above only yields the probability of observing the precise frequencies, in the expression, under the null. Of greater practical interest would be the specification of probabilities of *all* possible outcomes that are deviant from the null. That is to say, if the null hypothesis were true, then to assess the likelihood that resultant frequencies could have materialized under the null, at least the tail (or tails) of the multinomial sampling distribution must be constructed. Equation (2.7) would see repeated use during this most tedious, although possible, task. However, Eq. (2.7) need not be used to construct sampling distributions when, and if, the size of the sample is reasonably large. Instead, we can use the chi-square, for as we have learned this distribution approximates the multinomial well when the rule of five is satisfied.

THE PRODUCT-MULTINOMIAL RULE

An alternative to fixing the overall sample size (i.e., n), is to fix the size of respective levels of one (or more) of the variables. Recall that if one of the variables is an explanatory variable (as opposed to a response variable) then it is often desirable to stratify the population on the basis of the explanatory variable and to randomly draw a desired number of subjects from each population stratum. If, for example, the study dealing with attitude toward an antiabortion amendment were to be implemented at a military college where female students were decidedly in the minority, and if the investigator's intent were to compare male and female response to the abortion issue, then intentionally fixing the numbers of males and females in the sample so that both sexes will be well represented would be justifiable. In fact, if the total size of the sample is to be 100, then greatest statistical efficiency (power) for comparisons between females and males will result by drawing at random 50 females from the short roster and 50 males from the longer roster. But realize that arbitrarily fixing the sample sizes of the levels of the sex variables consequently precludes the opportunity to conduct a symmetrical analysis. Relationships apparently discovered in such an analysis would lack validity for the population of interest because the student sample would not be representative of the student population in the college. On the other hand, the fixed sampling scheme suggested above is valid for asymmetrical inquiry, inquiry that seeks to document differences between the proportional response of females and males with respect to the attitudinal response variable.

Now when the n's of respective levels of an explanatory variable are fixed, the exact sampling distribution of observed frequencies is given by a variant of Eq. (2.7) called the product-multinomial rule. To apply the product-multinomial to our working example, the following three operations will be put into effect:

1. The multinomial law (i.e., Eq. 2.7) will be used at the first level of the explanatory variable (i.e., $A_1 =$ females) to calculate the probability of observing f_{11} and f_{12} under the hypothesis that the respective population parameters are P_{11} and P_{12}.

2. The multinomial law will be used at the second level of the A variable (i.e., $A_2 =$ males) to assess the probability of realizing f_{21} and f_{22} under the hypothesized parameters P_{21} and P_{22}.

3. The product of probabilities calculated within levels of the explanatory variable will be taken as the probability of obtaining the four observed elementary cell frequencies in the fourfold table, under the known or assumed parameters.

Perhaps the product-multinomial is best appreciated through a careful study of its algebraic formulation. Stipulating that the sample size associated with the ith level of explanatory variable A will be symbolized by n_i^a (where in this instance, n_1^a equals the number of females and n_2^a equals the number of males), the product-multinomial equation for a fourfold situation would be

$$P(f_{11}, f_{12}, f_{21}, f_{22} | n_1^a, n_2^a) = \prod_{i=1}^{2} \frac{n_i!}{(f_{i1}!)(f_{i2}!)} P_{i1}^{f_{i1}} P_{i2}^{f_{i2}} \qquad (2.8)$$

As a careful examination of Eq. (2.8) reveals, differences in the number of subjects in levels of the sex variable (Variable A) do not systematically affect the product-multinomial probability. Incidentally, this is not true for Eq. (2.7). But for the product-multinomial, differences in the numbers of male and female participants are, in effect, "adjusted" so that only their respective profiles of proportional response to the attitudinal variable will affect the product solution. Thus it follows that it is the product-multinomial law that governs probabilities when there are one or more explanatory variables and when an asymmetrical analysis is desired. But again, in practice, exact sampling distributions prescribed by this law are rarely calculated. Instead, exact distributions are approximated by the chi-square when samples are deemed sufficiently large.

A summary may prove to be of value at this point. In this chapter we have learned that when two or more qualitative variables are crossed so as to yield cross-tabulations, the multinomial and product-multinomial laws describe the sampling distributions. The multinomial is assumed for cross-classifications resulting from a fixed sample of size n; the product-multinomial is assumed when the respective categories of at least one qualitative variable are determined or fixed by the investigator. The sampling model not only determines what manner of sampling distribution is appropriate but can also influence the "directionality" of an analysis. Simply put, random sampling from the population of interest is the requisite for a symmetrical analysis where exact probabilities under a specified null hypothesis may be generated by the multinomial law. Either a random or fixed sampling scheme, however, can be used for an asymmetrical analysis; but if the latter sampling scheme is selected, then it is the product-multinomial law that yields exact probabilities under the null. In the following chapters, both sampling distributions will be approximated by the chi-square in our work with larger samples. Moreover, it will be seen that the mechanics of analysis will not be affected by choice of sampling model, but the directionality of the analysis (symmetric vs. asymmetric)—and hence, the interpretation of results—will be very much tied to the manner in which the sample is drawn.

MAXIMUM–LIKELIHOOD ESTIMATION

The log-linear literature in general is redolent with cursory references to estimators and estimates derived by maximum-likelihood methods, an approach to parameter estimation developed in the 1920s by

R. A. Fisher. Unfortunately, it is taken for granted in most discussions of log-linear analysis that we are all equally familiar with the two most frequently used methods of parameter estimation: ordinary least squares (OLS) and maximum likelihood (ML). Parenthetically, both have been used in the analysis of complex contingency-table data; however, after a brief period of experimentation with least squares (see Grizzle et al., 1969; Theil, 1970), whereupon it was noted that estimates obtained by least squares "have a somewhat larger variance than maximum-likelihood estimates" (Goodman, 1972b, p. 45), estimation based on principles of maximum likelihood has emerged as the prominent procedure when working with models that attempt to explain qualitative response. However, many behavioral researchers trained in areas such as psychology, political science, and education are not as familiar with maximum-likelihood estimation as they are with least squares. Hence, we have included this section to point out the most basic features of this manner of obtaining estimates so as to reduce slightly the mystery in still one more area of log-linear work.

Desirable Properties of Estimators

We start by considering a *point estimator*, an inferential statistic that provides us with a numerical *estimate* of a parameter of interest. An arithmetic mean ($\bar{X}$), when computed on a sample so as to serve as an estimate of a population mean (μ), is a point estimator. An observed sample proportion (e.g., p_{ij}^{ab}) is a point estimator when used to estimate the population proportion P_{ij}^{ab}. To enhance our ability to generalize, let $\hat{\theta}$ stand for an unspecified estimator of parameter θ.

In statistics there are a number of procedural strategies that can be used to obtain a point estimate of θ. (OLS and ML are but two such methods.) As it happens, often two or three estimates, each emanating from a different method, stand in competition for the part of being chosen the "best" estimator of θ. The mean, median, and mode, for example, are estimators of μ. Selecting the so-called best estimator is not as difficult as it might at first appear, for over the years a number of desirable characteristics for estimators have been identified that collectively serve as criteria to aid in choosing between and among alternative statistics. In brief, these desirable characteristics for a good point estimator are:

1. *Unbiasedness.* A point estimator is unbiased if its *expected value* (the mean of its sampling distribution) is equal to the population

value to be estimated. In general, lack of bias means

$$E(\hat{\theta}) = \theta$$

which, loosely described, means that the sampling distribution of $\hat{\theta}$ centers about θ in the long run.

2. *Consistency.* To the extent to which an estimator approaches the value of the estimated parameter as the size of the sample becomes large, the estimator is said to be consistent. That is,

$$P(\hat{\theta} \to \theta) \to 1 \qquad \text{as } n \to \infty$$

which is read: the probability that $\hat{\theta}$ approaches θ as n becomes larger in size is unity. This means that as n becomes larger, the sampling distribution of the $\hat{\theta}$'s becomes more concentrated, less variable.

3. *Efficiency.* To the extent to which the sampling distribution of an estimator becomes concentrated, less variable, the estimator is efficient. Relatively efficient estimators have relatively small standard errors. That is,

$$\text{S.E.}(\hat{\theta}) \to \text{small}$$

Furthermore, comparatively speaking, the estimator whose sampling distribution has the smallest standard deviation (i.e., standard error) is sometimes called a *minimum variance estimator.*

4. *Sufficiency.* The property of sufficiency, attributed to R. A. Fisher, refers to the amount of sample information needed to estimate a parameter. An estimator is said to be sufficient if it utilizes all available sample information that potentially can enhance estimation of θ. Thus, $\hat{\theta}$ would be a sufficient estimator of θ if no other estimator can add to or improve upon the estimation of θ.

The Method of Maximum Likelihood (ML)

We will next attempt to grasp the essential features of one of the two main methods used to find estimators that possess one or more of the properties just outlined. The method to be considered is maximum likelihood—the more familiar method of least squares is discussed in many basic books on behavioral statistics. Following this consideration, we will point out some of the advantages of maximum likelihood for our work with contingency table data.

Much unlike least-square estimation, the mathematics of ML estimation necessitate a priori knowledge of how the sample event (statistic) of interest is distributed. Thus, if we desire to determine θ_{ML}, the maximum-likelihood estimator of θ, we must first be able to specify the *density function* (a mathematical equation describing a distribution) for the observed sample event. To illustrate, consider a sampling event whose density function is by now well known to us—the number of designated dichotomous events (say "successes," as opposed to "failures") observed over n trials or occasions, a statistic described by a density function known as the binomial law [Eq. (2.5)]. With this requisite knowledge, we write the function

$$P(f \mid n, P) = \binom{n}{f} (P)^f (1-P)^{n-f}$$

which, subsequent to study, is found to be a slightly more succinct, but equivalent, restatement of the binomial law of Eq. (2.5). This function can be read as follows: the probability of observing a number of "successes" equal to f in a sample of size n is, in addition to n, a function of P where P represents the proportion of "successes" in the population from which the sample was drawn. Recall that if we know (or are willing to assume) the numerical value of the parameter P, it can be inserted in the function and the probabilities of various sample outcomes (f's) can then be determined.

But our intent is not to use the density function to generate outcome probabilities or to construct a sampling distribution. Instead, in pursuit of solving for the ML estimator, we must reverse our line of reasoning by starting with a sample outcome (say $f = 9$ in a sample where $n = 15$) and then asking the question: What value of P would make $f = 9$ most likely? The central idea underlying the identification of the ML estimator for situations in general, and a situation of dichotomies in particular, is to search among the many possible values that can be taken on by P and to find the value that is most likely for a given value of f. The most-likely parameter (P) is the maximum-likelihood estimator of P.

The mathematical identification of most-likely parameters given certain sample results requires the use of differential calculus. Without going into detail, essentially, we would differentiate the density function with respect to the unknown most-likely parameter, set the resultant derivative to zero, and find the maxima.[2] If we use

---

[2]Should a reader with command of the calculus attempt to do this on the density function for the binomial, computations will be easier if the multiplicative function is transformed to a linear function by taking logarithms of respective factors.

the suggested calculus on the density function for the binomial distribution, the most likely value of P given an observed outcome f conveniently turns out to be f/n. Simply put, we can say that the ML estimator of a population proportion turns out to be the statistic p, the corresponding proportion observed on a representative sample. Most readers will have known intuitively that $p = \hat{P}_{ML}$; that for $f = 9$ and $n = 15$, $\hat{P}_{ML} = 9/15 = .60$. No other value of P (e.g., .58 or .62) would be as likely.

We have established that there are four main properties that make for a good estimator and that there are a number of statistical approaches to the generation of parameter estimators, and from this the reader may have hypothesized that different methods of estimation tend to promote the attainment of different properties. Indeed they do. And although estimators produced by the method of least squares have much to commend them—least squares still has no rival when it comes to fitting curves to data when certain assumptions about that data are tenable—ML estimators, when obtainable, are usually extremely attractive. This is because ML estimators are both consistent and efficient, at *least* as consistent and efficient as any other estimator, and ML estimators can be said to be sufficient if a sufficient estimator, in fact, exists. It is only in one respect that ML estimators fall a bit short of the mark: more often than not, these tight and tidy estimators tend, unfortunately, to possess some degree of biasedness.

This review of ML estimation has been provided for good reason—ML estimation is used extensively in the approach to log-linear analysis taken in this book. It is not difficult to appreciate the value of ML methods for work with qualitative data. After all, ultimately we will be concerned with estimating the proportion of subjects belonging to certain variable combinations and using these estimates either to identify relations or to document group difference. And since we know that sample proportions follow the multinomial law, that sample proportions are ML point estimators of population proportions, that these estimators are maximally consistent, decidedly sufficient, and relatively efficient, it is not surprising that both maximum-likelihood point and interval estimation generally are preferred over least squares for the work to be described in the chapters that follow.

3

Traditional Applications
of Chi-square

Since its invention in 1900 by Karl Pearson, the use of the χ^2 statistic in behavioral research has been extensive. A χ^2 statistic, for example, is commonly used when the variable or variables of interest are qualitative and a test of the agreement is desired between *observed* frequencies and the frequencies to be *expected* (theoretical frequencies) on the basis of some hypothesis. Because the distribution of the χ^2 is so basic to theory in statistics, and because the χ^2 statistic is so versatile in practical application, all who work directly with behavioral data or make decisions based upon such data should be familiar with its theory and use. But unfortunately, as has been stated by Lohnes and Cooley (1968), "χ^2 theory is as elusive as its applications are pervasive" (p. 145). Hence, this chapter: a chapter designed to review the elemental features of χ^2 theory, to define conditions in which the use of χ^2 is appropriate, and to illustrate the computations associated with its traditional applications. After all, knowledge of the chi-square is a sine qua non to the study of log-linear analysis.

THE χ^2 AND ITS SAMPLING DISTRIBUTION

We begin by examining the χ^2 statistic somewhat removed from practical application. Hence, consider an infinitely large *population* of scores ($N \to \infty$) that are distributed normally and independently.

An unspecified score in this population will be denoted by X_i, where $i = 1, 2, \ldots, N$. The *variance* of population X_i's, a parameter, is represented by σ^2.

Now assume that a *sample* of size n is randomly drawn from the foregoing population. From the sample the arithmetic mean $(\bar{X})$ and an unbiased variance estimator can be readily calculated. The estimator of variance, which is *not* a maximum-likelihood (ML) estimator in this case, is

$$\hat{\sigma}^2 = \frac{\sum_i^n (X_i - \bar{X})^2}{n - 1} \tag{3.1}$$

If the population variance is either known or is assumed known under a null hypothesis, the following statistic is amenable to calculation:

$$\chi^2 = \sum_i^n z_i^2 = \sum_i^n \frac{(X_i - \bar{X})^2}{\sigma^2} = \frac{(n-1)\hat{\sigma}^2}{\sigma^2} \tag{3.2}$$

This statistic, verbally defined as the sum of n squared normal deviates, is an *exact* χ^2 statistic—a statistic whose probability distributions are well known.

In a manner similar to the t statistic, the probability distribution of χ^2 is dependent upon sample size. That is, the shape, variability, and mean (i.e., expected value) of sampling distributions of χ^2's depend on the number of degrees of freedom associated with the sample. To achieve economy of expression, we will often use ν as a general symbol for the number of degrees of freedom. Thus, with respect to Eq. (3.2) and a sample of size n, $\nu = n - 1$.[1]

The sampling distributions of χ^2's are most interesting because they are solely determined by their number of degrees of freedom. For example, the expected value (mean) of a distribution of χ^2 is the numeric value of its degrees of freedom. That is,

---

[1] The number of degrees of freedom is equivalent to $n - 1$ because, in the execution of Eq. (3.2), the sample mean $(\bar{X})$ is used as an estimator of the population mean (μ) which need not be known. Since the sample mean has been "fixed" in its role as a parameter estimate, only $n - 1$ sample values are free to vary and hence contribute to the variance estimate.

$$E(\chi^2) = \nu$$

The *standard error* (standard deviation) of a χ^2 distribution is also a function of ν. Specifically, it is

$$\text{S.E.}(\chi^2) = (2\nu)^{1/2} = \sqrt{2\nu}$$

With respect to the underlying metric of these distributions, the lower bound is zero since the numerical value of the χ^2 cannot be negative and the upper bound extends to infinity. In addition, χ^2 distributions are nonsymmetrical, except for large *n*'s. Sketches of several χ^2 distributions are presented in Figure 3.1.

Finally, the reader should know that the exact χ^2 given by Eq. (3.2) has important applications for response measures that are interval or ratio in nature. For example, Eq. (3.2) is used to test a null hypothesis that the population variance (σ^2) is equal to a specified value (see Hays, 1973, pp. 439–441). Also, the χ^2 plays an important role in multivariate analysis, for the sampling distributions of multivariate normal distributions can be shown to be a function of the χ^2 (see Tatsuoka, 1971, pp. 67–73). But our present concern is not with the use of the χ^2 for interval data but rather with how χ^2 procedures are used with qualitative variables and resultant qualitative data.

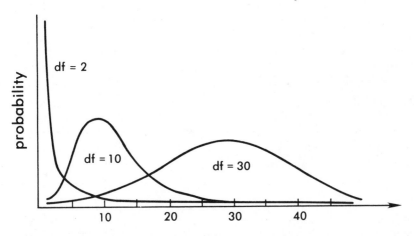

Figure 3.1. Chi-square distributions associated with 2, 10, and 30 degrees of freedom.

APPLICATIONS FOR QUALITATIVE DATA

There are at least three research situations in which the chi-square can assist the researcher with the analysis of qualitative data. Each is defined by a different number of variables. These situations are:

1. There is one qualitative variable and the researcher desires to compare the number of cases that are *observed* to fall into specific categories with the number of cases that are *expected* to fall into each category on the basis of some hypothesis. The situation calls for a one-variable test of *goodness-of-fit.*

2. There are two crossed categorical variables and the researcher desires:

(a) either to determine if there is a significant association, or relationship, between the two variables, a situation which calls for a test on a null hypothesis of variable *independence* (i.e., a symmetrical analysis).

(b) or to determine if the pattern of response to one variable is similar over all categories of the second variable, a situation which calls for a test on a null hypothesis of *homogeneity of response* (i.e., an asymmetrical analysis).

3. There are more than two crossed qualitative variables and the intent is either to assess associations among these variables or to determine if there are differences in the pattern of responding to one of these variables. Here, chi-square goodness-of-fit procedures are used, but within the system of analysis called log-linear contingency table analysis.

In this chapter, we limit discussion to the first two situations above; log-linear analysis will be introduced in the next chapter. With respect to the first two situations, we learned from Chapter 2 that exact sampling distributions are given by the binomial, multinomial, or product-multinomial laws. However, the actual use of equations given in Chapter 2 to construct these exact sampling distributions, or even the tails of these distributions, will prove to be involved and formidable, especially for larger values of n. But when n's are sufficiently large—say, large enough to satisfy the rule of five—the construction of exact sampling distributions is not necessary, for we also learned that the convenient and highly accessible distribution of the chi-square can be used as an approximation to the exact distributions. But to use the χ^2 distribution as a proxy, a way must be found to compute a χ^2 statistic (or, an approximate χ^2 statistic) on qualitative data—not on interval data as is assumed by Eq. (3.2).

Fortunately, in addition to developing the *parametric* statistic given by Eq. (3.2), Pearson also discovered that under certain circumstances the χ^2 statistic (hence its distribution) can be approximated with qualitative data. To be specific, at the turn of this century Pearson introduced a *nonparametric* version of the chi-square—a version that involved fewer assumptions relating to the parameters μ and σ^2—which, if n is sufficiently large, is approximately distributed as a χ^2. And since the discrete multinomial approaches the continuous χ^2 as samples become large, Pearson's nonparametric statistic which approximates the continuous χ^2 represented a welcomed solution to many problems formerly encountered when analyzing qualitative data.

The Pearson approximate χ^2 uses observed sample frequencies within categories of qualitative variables (i.e., np_i's or f_i's) and systematically compares these frequencies to frequencies that would be expected under a null population with population proportions $P_1, P_2, \ldots, P_k$. This classical approximation is given by

$$\chi^2 = \sum_i^k \frac{(np_i - nP_i)^2}{nP_i} = \sum_i^k \frac{(f_i - F_i)^2}{F_i} \tag{3.3}$$

To repeat, the nonparametric statistic given by Eq. (3.3) is approximately distributed as a χ^2 distribution possessing ν degrees of freedom.

Although the derivation of the Pearsonian χ^2 approximation is well beyond the scope of most books, including this one, the nature of its derivation is nonetheless intuitively pleasing. Notice, for instance, that for each category of a qualitative variable, we obtain first the difference between the observed frequency and that which is expected for that category assuming that the population proportion was P_i. That is, for each of the k classes or categories, we compute

$$f_i - F_i$$

Now if we were to simply sum these differences over categories, the sum would turn out to be zero since

$$\sum_i^k (f_i - F_i) = \sum_i^k f_i - \sum_i^k F_i = n - n = 0$$

However, by squaring each observed/expected difference prior to summing, this problem can be circumvented. But before squared differences are summed, notice that each is divided by its *expected* frequency. By so doing, the squared discrepancy for each category is weighted by F_i prior to summation. Even this weighting operation makes sense if one considers that a given discrepancy in a category in which few observations are expected (i.e., F_i is small) is more indicative of departure from a hypothesized population than it would be if it were associated with a category with a large expected frequency. Finally, note that the sum of weighted squared observed/expected differences constitutes the test statistic, a statistic which we have said is approximately distributed as a χ^2 possessing a number of degrees of freedom equal to the number of restrictions placed on sample calculations. As we shall see, in situations involving only one qualitative variable the number of restrictions is generally one, and hence most often $\nu = k - 1$.[2]

To ensure their proper distribution, all statistics, even nonparametric statistics, should in theory satisfy certain assumptions and meet certain conditions. Thus, to ensure that the nonparametric Pearsonian statistic defined by Eq. (3.3) does, in fact, approximate the multinomial, three general conditions relating to methods of data collection should be met. These general conditions are:

1. Members of the sample who are to provide data should be independently sampled with equal probability from the population of interest. In short, in theory, the validity of our inference is based on the assumption that simple random sampling was used for a symmetrical analysis and either simple or stratified random sampling was used for an asymmetrical analysis. Obviously, often in practice this condition cannot be satisfied completely.

2. Members of the sample should be classified into categories of each and every variable such that the classification process can be said to be *independent, mutually exclusive*, and *exhaustive*. The first is important because distributions described by the multinomial law assume that each and every observation is independent of each other

---

[2] The calculation of the chi-square statistic in typical one-factor situations is restrained by the fact that the sum of F_i's equals n. Consequently, if $k - 1$ expected frequencies are known, the remaining frequency can be determined. There is, however, a special case, a goodness-of-fit test for a normal population where, in addition to the above restraints, two parameters (μ and σ) have to be estimated to perform the test. For this special case, $\nu = k - 3$.

observation. The consequence of compliance with the independence of response condition is that members of the sample can only be assigned to one category of a variable; therefore, it follows that members can be placed into only one cell of a contingency table. The direct incorporation of a repeated measurements variable into a contingency table, such as is routinely done in mixed-variable ANOVA designs, is not appropriate here. The second member of the trio, mutual exclusivity, means that each variable should be structured and defined so that members drawn from the population are assigned to one and only one category. Finally, exhaustiveness implies that each and every variable has categories sufficient to accommodate all members of the population as defined by the researcher. Again, all three specific conditions are needed to satisfy the multinomial that the χ^2 attempts to approximate.

3. Members of the sample should be present in sufficient numbers (the n should be sufficiently large) so that exact multinomial probability distributions approach exact χ^2 distributions (under the central limit theorem) and, in turn, the distributions of Pearsonian χ^2's approach exact χ^2's. For the binomial situation, it has been established that approximate methods can be used when the *rule of five* is satisfied. As a rule, this rule is also applicable to polytomies and multifactor contingency tables. When generalized to these situations, the rule of five can be taken to mean that the size of the sample should be large enough to ensure that no category (or cell) contains an *expected* frequency (not an observed frequency) of less than five cases. Applying the χ^2 to tables in which expected frequencies fall below the level of five is thought by many to produce χ^2 values that are spuriously large, especially if the table is a fourfold table. For larger tables, however, it appears that the rule of five can be relaxed somewhat without serious inflation of the Type I error rate (cf. Hays, 1973, p. 736). One approach to this problem has been to combine neighboring categories so as to satisfy the rule, but this remedy must be implemented with caution for not only can combining categories result in a loss of statistical power (Cochran, 1954) but, as we shall see, it can result in a marked distortion of findings.

With this much background, we are now in the position to appreciate the applications of the chi-square in behavioral research. The first illustration involves only one variable and the use of the Pearsonian χ^2 to perform a goodness-of-fit test. Our attention will then be directed to several multi-variable situations to witness both a test of independence and a test of homogeneity of response.

SINGLE VARIABLE GOODNESS-OF-FIT

Essentially, tests of this genre attempt to determine whether a distribution of observed sample frequencies is sufficiently well-fitted, or compatible, to some theoretical form. The theoretical form will be either a binomial or a multinomial population with specific parameters (P_i's), parameters that are specified by the researcher when he or she posits a null hypothesis. Through the use of Eq. (3.3), observed sample frequencies (f_i's) are compared to frequencies to be expected under the null (F_i's), a χ^2 statistic is computed, and the statistic is related to a tabled distribution of χ^2's to determine the likelihood of drawing the observed sample distribution from the population specified by the null hypothesis.

Consider the following simple example. Assume that the dean of a college of social work believes that graduate students are evenly divided with respect to their opinions on requiring statistics courses in their programs. To subject this belief to test, the dean plans to obtain a modest but representative sample of graduate students (say $n = 30$) and poll them on this matter. Subsequent to polling, students will be classified into one of three mutually exclusive categories (i.e., $k = 3$). These categories are: (a) in favor of requiring statistics, (b) opposed to requiring statistics, and (c) undecided.

The dean's belief concerning an equal division of opinion can be translated into the following null hypothesis:

$$H_o: P_1 = P_2 = P_3 = .33$$

where P_1 symbolizes the proportion of students in the population who favor statistics as a requirement, P_2 the proportion opposed, and P_3 the proportion undecided.

Turning to the sample, assume that 16 students favored a statistical requirement, 9 indicated that they were opposed to such a requirement, and 5 were judged to be undecided on this matter. These values constitute observed frequencies (f_i's) within levels of the qualitative opinion variable.

It is, however, the researcher's task to generate the *expected* frequencies for each level based on the null hypothesis. Since $P_1 = .33$, the expected number of students in a sample size 30 that would favor a statistical requirement is $F_1 = nP_1 = (30)(.33) = 10$. By similar logic, F_2 and F_3 also would be 10 in this particular case. A working table for the goodness-of-fit test is provided by Table 3.1.

TABLE 3.1. Differences between Observed and Expected Frequencies by Expressed Opinion on Statistics as a Requirement

Opinion Variable	f_i	F_i	$(f_i - F_i)^2/F_i$
In Favor	16	10	3.60
Opposed	9	10	.10
Undecided	5	10	2.50
Totals	30	30	6.20

Note. Expected frequencies are those expected under the null H_0: $P_1 = P_2 = P_3 = .33$.

Rewriting Eq. (3.3) and substituting into it values appropriate to the problem yields

$$\chi^2 = \sum_i^k \frac{(f_i - F_i)^2}{F_i}$$

$$= \frac{(16-10)^2}{10} + \frac{(9-10)^2}{10} + \frac{(5-10)^2}{10}$$

$$= 3.60 + .10 + 2.50 = 6.20$$

To complete the test, the computed Pearsonian χ^2 statistic is related to a χ^2 distribution possessing $k - 1$ (i.e., 2) degrees of freedom. Consulting a table of critical values for the chi-square distribution (given in Appendix B) reveals that a test statistic equal to or in excess of 5.99 is needed to be able to claim statistical significance at the .05 level. Since $\chi^2(2) = 6.20$, significance can be claimed.

The dean, therefore, can conclude that the null hypothesis of uniform distribution of opinion on this issue is not tenable. Moreover, for a simple problem like this, it is a relatively simple matter to identify the sample categories that are most deviant from those posited under the null. From an examination of the respective contribution made by each category of opinion to the numerical value of the χ^2 statistic (e.g., contributions were 3.60, .10, and 2.50, respectively) it is evident that the sample distribution does not depart from the hypothesized distribution in the "opposition" category, but rather that the "poorness-of-fit" is due to a greater than expected number of students with "favorable" opinions and a fewer than expected number of students who were classified as "undecided." There are still other techniques, collectively called *residual analyses*, that can be of help when attempting to identify categories (or cells) that deviate markedly from the null distribution. Several of these techniques will be introduced in a slightly more challenging context later in this chapter.

To illustrate the point that the χ^2 test can be very sensitive to grouping, let us assume that the dean decided to redefine the categorical opinion variable. That is, instead of a trichotomy, the opinion variable is now treated as a dichotomy consisting of: (1) students in favor of requiring statistics and (b) students not expressly in favor of requiring statistics. To accommodate the new second category, the second and third categories of the previous example will be combined. The null hypothesis of uniform distribution is now $H_o:P_1 = P_2 = .50$. Regrouping the observed frequencies given in Table 3.1, which results in 16 students in favor of statistics and 14 students not in favor of statistics as a requirement, and performing the single degree of freedom χ^2 test gives $\chi^2(1) = .13$, a result that the reader is encouraged to verify. (Since $k = 2$ here, a z test such as was illustrated in Chapter 2 could have been performed with comparable results.) Because the computed test statistic falls far short of $\chi_\alpha^2 = 3.84$, the tabled critical value at the .05 level, there is no basis for rejecting the null hypothesis. Thus, unless there is a strong theoretical justification for a certain grouping procedure, be careful, for as we have seen arbitrary classification can affect results.

THE HYPOTHESIS OF VARIABLE INDEPENDENCE

The χ^2 test performed on a null hypothesis of variable independence is essentially an extension of the previous test of goodness-of-fit. For the case of two qualitative variables, the researcher: (a) creates a hypothesis (i.e., a null hypothesis) which reflects a condition of variable independence between the two variables, (b) draws a sample of n members and jointly classifies members on the basis of the two variables, and (c) compares observed cell frequencies with those expected under the null for goodness-of-fit.

Independence in Fourfold Tables

In Chapter 1, two dichotomies were crossed to form a fourfold table. Recall that subjects were college students jointly classified by sex and by attitude toward an antiabortion amendment, henceforth referred to simply as the "Attitude" variable. Let our mode of inquiry by symmetrical. That is, we desire to determine whether a statistically significant relationship (or association) exists between Sex and Attitude—suggesting that if one had knowledge of a student's sex, one would be able to predict his or her attitude relative to the abortion issue better than would be expected by chance, or vice versa. Before proceeding further, a review of the contents of Table 1.1 is suggested.

We formally begin by advancing the null hypothesis that dichotomous Variables A (Sex) and B (Attitude) are independent. Symbolically, this may be expressed as:

$$H_o : A \otimes B$$

It should be understood that this hypothesis relates to a *population* of college students who in this instance belong to one—and only one—of the following four variable combinations:

AB_{11} = females opposed to the amendment

AB_{12} = females in support of the amendment

AB_{21} = males opposed to the amendment

AB_{22} = males in support of the amendment

Now if the null hypothesis is precisely true and if two parameters are known, the exact *proportion* of students belonging to each of the four cells or variable combinations above can be determined. The first parameter to know is the proportional representation of females (or males) in the population, i.e., P_1^a. The value of this parameter, of course, gives us the value of P_2^a, the proportion of males in the population. Second, we need to know the proportion of subjects in the population that fall into one of the categories of the Attitude variable, say P_1^b. With knowledge of these *marginal* parameters, *if there is no relationship* between Sex and Attitude, then the proportion of students belonging to the cell located in the ith row of A and the jth column of B is

$$P_{ij}^{ab} = P_i^a P_j^b \tag{3.5}$$

To illustrate this most important multiplicative principle, suppose that 50% of population members were females (i.e., $P_1^a = .50$) and it turned out that 60% of population members opposed the antiabortion amendment (i.e., $P_1^b = .60$). If Sex and Attitude are in fact independent, then the proportion of females who oppose the amendment would be

$$P_{11}^{ab} = P_1^a P_1^b = (.50)(.60) = .30$$

Moreover, deviation from $P_{11}^{ab} = .30$ in this instance constitutes evidence in support of a relationship or association between the two variables, for knowledge of a student's sex would permit a better-than-chance prediction of that student's attitude toward abortion, or vice versa.

Thus far, we have the identity given by Eq. (3.5) that represents, in effect, the state of variable independence. The identity relates to a population, yet in practical research work we rarely have access to populations and therefore rarely know the numerical value of parameters. Instead, we try to obtain samples that are representative of populations and to estimate parameters using statistics. For purposes of this example, therefore, assume as before that a sample of 100 college students has been randomly drawn and jointly classified by Sex and Attitude as indicated by Table 1.1.

The sample information provided in Table 1.1 will now be used to: (a) *estimate* the marginal parameters P_i^a and P_j^b shown to the right of the equals sign in Eq. (3.5) and, in turn, (b) to *estimate* P_{ij}^{ab}, the expected proportion (or probability) for the ijth cell. Recall from the last chapter that the ML estimators of parameters P_i^a and P_j^b are sample p_i^a and p_j^b, respectively. Moreover, it turns out that these ML estimators are unbiased (which often is not true of ML estimators), so we can say that $E(p_i^a) = P_i^a$ and $E(p_j^b) = P_j^b$.

Having found highly suitable estimators of the marginal parameters, we turn to the estimation of P_{ij}^{ab} which, under the null hypothesis of *independence*, is

$$\hat{P}_{ij} = p_i^a p_j^b \tag{3.6}$$

where the carat or "hat" (^) over P_{ij} will again be used to designate a population estimator. But since the conventional formula for the approximate χ^2 uses frequencies instead of proportions, we usually work with *expected frequencies* for the cells of the contingency table. These are given by

$$F_{ij} = n\hat{P}_{ij} = np_i^a p_j^b$$
$$= (n)(n_i^a/n)(n_j^b/n)$$
$$= \frac{n_i^a n_j^b}{n} \tag{3.7}$$

For example, the expected frequency for females who are opposed to the antiabortion amendment is

$$F_{11} = \frac{n_i^a n_i^b}{n} = \frac{(40)(70)}{100} = 28$$

Only one F_{ij} need be computed by formula because, in a fourfold table, if one F_{ij} is known the remaining three may be obtained by subtraction from respective fixed marginal frequencies.

Our work to this point is summarized in Table 3.2, where we note that expected cell frequencies are contained within parentheses and that the size of the sample is sufficient to accommodate the chi-square approximation to the multinomial.

To apply Pearson's approximate chi-square to a contingency table, weighted-squared-cell differences are summed over rows ($i = 1, 2, \ldots, a$) and columns ($j = 1, 2, \ldots, b$). With this in mind, we compute

$$\chi^2 = \sum_{ij}^{ab} \frac{(f_{ij} - F_{ij})^2}{F_{ij}} \tag{3.8}$$

$$= \frac{(33-28)^2}{28} + \frac{(7-12)^2}{12} + \frac{(37-42)^2}{42} + \frac{(23-18)^2}{18}$$

$$= .8929 + 2.0833 + .5952 + 1.3889$$

$$= 4.96$$

In general, the number of degrees of freedom that prescribe the relevant χ^2 sampling distribution for contingency tables is the product $(a-1)(b-1)$. Therefore, for fourfold contingency tables, $\nu = (2-1)(2-1) = 1$. Relating the computed χ^2 statistic to the tabled critical value associated with the .05 level of significance, and $\nu = 1$, reveals that $\chi^2(1) = 4.96 > \chi_\alpha^2 = 3.84$; hence, rejection of the independence hypothesis is justified. We find that there is a statistically significant association between the two dichotomous variables.

TABLE 3.2. Observed and Expected Frequencies for the Hypothetical Study Introduced in Chapter 1

Sex of Subjects	Attitude toward the Amendment		Marginals
	Opposed	Support	
Females	33 (28)	7 (12)	40
Males	37 (42)	23 (18)	60
Marginals	70	30	100

Note. Expected or theoretical cell frequencies are contained within parentheses.

Usually, however, the mere documentation of an association does not complete the analysis. Questions concerning the *directionality* and *intensity* of the association remain. As will become apparent later in this chapter, for large, complicated tables a residual analysis can prove to be of great assistance in determining the directional nature of an association. For relatively simple designs, an examination of differences between observed and expected cell frequencies (see Table 3.2) will usually reveal the essential direction of the association. For the present example, such an examination reveals that males at this college are more supportive of the antiabortion amendment than are females. The strength or intensity of the association, however, is a separate matter. As a reading of the section on follow-up procedures in Chapter 4 will disclose, Yule's *Q* statistic would be a highly appropriate measure of the strength of the relationship in this fourfold table.

Independence in Larger Tables

As long as the previously discussed requisite conditions are satisfied, the test of variable independence can be extended to contingency tables of order greater than 2 × 2. Such a table has been constructed by this writer from data reported by O'Connor and Sitkei (1975). Briefly, these investigators conducted a national survey of community-based residential homes for retarded persons "to provide an initial profile of these facilities and their residents" (p. 35). One aspect of their research centered about the relationship between the Type of Facility (public, private nonprofit, or private profit) and the Number of Residents served by the facility (up to 10, between 11 and 20, or

TABLE 3.3. Cross-Classification of Facilities by Type and the Number of Retarded Residents Served

Type of Facility	Number of Residents (#R)			
	#R $\leqslant$ 10	10 $<$ #R $<$ 21	#R $\geqslant$ 21	Marginals
Public	67 (46)*	45 (48)	19 (37)	131
Nonprofit	98 (99)	98 (105)	88 (80)	284
Private Profit	48 (68)	83 (73)	65 (55)	196
Marginals	213	226	172	611

Note. This table has been constructed from data presented as bar graphs by O'Connor and Sitkei (1975) with permission of the publisher, the American Association on Mental Deficiency.

*Expected cell frequencies, rounded to the nearest whole number, are contained within parentheses.

over 20). The cross-classifications of the 611 institutions surveyed by O'Connor and Sitkei are presented in Table 3.3.

Although O'Connor and Sitkei did not subject their data to inferential analysis, we shall. Specifically, let us test the null hypothesis that there is no association between Type of Facility and the Number of Residents. To obtain expected cell frequencies under the assumed condition of variable independence, Eq. (3.7) is employed. The value of F_{11}, for example is

$$F_{11} = \frac{n_1^a n_1^b}{n} = \frac{(131)(213)}{611} = 45.67$$

We need to use Eq. (3.7) at least three more times because four expected cell frequencies must be known before the remaining F_{ij}'s can be obtained as residuals from appropriate marginals. (It follows, therefore, that $v = 4$ for this problem.) Using the F_{ij}'s given in Table 3.3, which have been rounded to the nearest whole number, Eq. (3.8) is used to compute the chi-square test statistic:

$$\chi^2 = \frac{(67-46)^2}{46} + \frac{(45-48)^2}{48} + \cdots + \frac{(65-55)^2}{55}$$

$$= 28.88$$

Realizing that $v = (a-1)(b-1) = (3-1)(3-1) = 4$, reference to tabled values of chi-square reveals that $\chi^2(4) = 28.88$, $p < .001$.

Evidence is sufficient to reject the null and conclude that there is an association between the two variables in question. Moreover, a preliminary study of discrepancies between observed and expected cell frequencies in Table 3.3 tends to support the conclusion advanced by O'Connor and Sitkei that "Public facilities tend to have the fewest number of residents and to use small facilities while private profit facilities appear to be slightly larger than the non-profit or public facilities" (p. 36). (The substantive implications of this finding were not addressed by the authors.) We will attempt a more thorough follow-up of this analysis in the forthcoming section devoted to residual analyses.

THE HYPOTHESIS OF HOMOGENEITY OF PROPORTIONS

Still another use of the χ^2 is to assess whether groups of subjects respond differently on a categorical variable. Consider, for example, an experiment that involved two groups: one that received a special treatment and one that served as a control group. At the end of the experiment, subjects in each group were administered a task and assigned a performance rating of either "pass" or "fail." Or consider a descriptive study which seeks to determine whether teachers, guidance counselors, and administrators are similar with respect to their desire to be represented by the NEA, by the AFT, or not by a union or professional organization at all. In this instance it is unlikely that the investigator is interested in the relationship between the group variable and the response variable; instead he or she probably wants to know whether or not the various groups (an explanatory variable) respond differently over levels of a response variable. More to the point, the investigator probably wants to know if the *proportions* associated with each class of the response variable are approximately the same for each of the groups. To partially answer this question, a χ^2 test can be performed on the hypothesis of homogeneous proportional response.

Often the similarities and differences between χ^2 tests of independence and homogeneity of proportions are not well appreciated. Indeed, both use the Pearsonian χ^2, assuming of course that basic requirements are met. Differences, however, may be seen in the (a) sampling plans, (b) initial interpretation of results, and (c) procedures used to follow-up an initial significant result.

Turning first to sampling, recall that only the overall sample size is fixed by the investigator for a test of independence. Theoretically, the investigator defines a population of interest, proceeds to obtain a representative sample of size n, and cross-classifies sample members on the basis of selected qualitative variables. In contrast, for tests of homogeneity of proportions, the investigator may arbitrarily fix the marginal frequencies of the explanatory variable. We learned, for example, that the investigator may wisely choose to make each category of the explanatory variable of equal size despite the fact that group n's may not be equal with respect to the background population. Having fixed the marginals of the explanatory variable, subjects within fixed groups are then classified according to their status on the response variable. Such a sampling scheme, of course, would vitiate the value of a test of independence, but since our present concern centers about differences and similarities in group response, here it is appropriate.

After observed and expected frequencies are established, a conventional Pearsonian χ^2 statistic is computed to test the null hypothesis of uniform proportional response. As a rule, the sampling distribution approximated by the Pearsonian χ^2 is given by the product-multinomial law, not the multinomial law which is the basis for tests of independence. Nevertheless, the computation of the χ^2 statistic is carried out in the usual way, but, as we shall see, interpretation of results is different.

An example of a study that sought to test a hypothesis of homogeneity of proportional response has been reported by Goldsmid et al. (1977). The study examined selected attributes of "superior" college teachers in a large university. Sixty faculty members who were either awarded the honor of being a distinguished teacher or who received a significant number of nominations for this award were identified. In addition, a group of 60 control members was randomly selected from the faculty at large. Hence, the explanatory variable, Type of Teacher, consisted of two levels: distinguished vs. control.

One of several hypotheses advanced in this study was that the quality of teaching will peak in mid-career. To determine if there was support for this hypothesis, teachers within each of the two levels of the explanatory variable were in turn classified on the basis of three levels of chronological age. The categories of the Age variable were: under 39, 40 to 49, and over 49.[3]

---

[3] In their report, Goldsmid et al. structured a four-level Age variable, a variable that contained an "under 30" category. But, because this latter category contained a total of only four faculty, with permission this writer has taken the liberty to combine these four youngsters with members of a group ranging in age from 31–39 to form the "under 39" group used in the present illustration.

Keep in mind that the purpose of this study was not to investigate a possible relationship between Age and Type of Teacher but rather to determine whether distinguished teachers differed from control teachers with respect to selected attributes such as age. To specifically accommodate this purpose, a sample drawn at random from the population of faculty members at this institution was not required. Instead, the faculty at large, in effect, was stratified on the basis of distinguished members and typical (control) members so that 60 of each type could be drawn, a sampling scheme that maximizes the sensitivity of comparisons between these two groups.

The frequencies of occurrence by levels of the explanatory variable and by the three levels of the response variable turned out as shown in Table 3.4. Note that in addition to f_{ij}'s, the table also contains the F_{ij}'s that would be expected under the hypothesis of mutual variable independence. (The F_{ij}'s were calculated by methods discussed in the previous section; and, even though we are not interested in variable independence per se, the F_{ij}'s will be shown to be relevant to the problem at hand.)

There is a perspective to be highly recommended for this and similar problems. It is useful to view each level of the explanatory variable (e.g., Type of Teacher) as an intact entity, an entity that may or may not contain the same number of subjects as other entities. This view readily accommodates the conversion of frequencies within these entities to proportions that sum to unity. That is, separately

TABLE 3.4. Cross-Classifications by Type of Teacher and Age

Type of Teacher	Age Categories (B)			Marginals
	$B < 40$	$40 \leqslant B \leqslant 49$	$B > 49$	
Distinguished	30 (29.0)*	20 (18.5)	10 (12.5)	60
Control	28 (29.0)	17 (18.5)	15 (12.5)	60
Marginals	58	37	25	120

Note. Data reproduced with minor modifications from Goldsmid, C. A., Gruber, J. E., and Wilson, E. K., Perceived attributes of superior teachers (PAST): An inquiry into the giving of teacher awards. *American Educational Research Journal*, 1977, *14*, 423–440. Copyright 1977, American Educational Research Association, Washington, D.C.

*Expected cell frequencies under the hypotheses of variable independence are contained in parentheses.

by levels of the explanatory variable, envision the standardization of response by transforming frequencies to proportions such that within levels the proportions sum to 1.00. *Standardization to proportional unity* within groups facilitates the comparison of profiles of response between groups. Should the profile of proportional response to the Age variable provided by distinguished teachers turn out to be similar to the profile provided by the controls, respective proportional response is said to be homogeneous. Should the profiles of proportions turn out to be decidedly different, however, the response profiles are said to be heterogeneous. Let us illustrate this with the data at hand.

Assuming Variable A to be explanatory, the profile of frequencies within a level of A can be standardized to proportional unity by

$$p_{ij} = f_{ij}/n_i^a \tag{3.9}$$

Thus, for distinguished teachers, observed proportions standardized to unity are:

$$p_{11} = 30/60 = .500$$
$$p_{12} = 20/60 = .333$$
$$p_{13} = 10/60 = .167$$

For the controls, parallel operations yield $p_{21} = .467$, $p_{22} = .283$, and $p_{23} = .250$.

Having established the pattern of *observed* response for each group of teachers, we next determine *expected* patterns of proportional response for each group, assuming that these patterns are homogeneous in the background populations. As the reader may have surmised, expected proportions standardized to unity within groups are a function of the F_{ij}'s in Table 3.4. They are given by

$$P_{ij} = F_{ij}/n_i^a \tag{3.10}$$

Thus, for distinguished teachers, expected proportions are:

$$P_{11} = 29.0/60 = .483$$
$$P_{12} = 18.5/60 = .308$$
$$P_{13} = 12.5/60 = .208$$

Parallel calculations performed on control teachers yield an identical profile. That is, under the hypothesis of homogeneity of response, $P_{21} = .483$, $P_{22} = .308$, and $P_{23} = .208$. Our work to this point is summarized in the Table 3.5.

Several inferences can be drawn from careful study of Tables 3.4 and 3.5. The first is that when the F_{ij}'s for a test of mutual independence are transformed to proportions within levels of an explanatory variable (such that they sum to unity), profiles of proportional response within levels of the explanatory variable will be identical. Hence, we conclude that the *expected distributions* for tests on the hypothesis of independence and tests on the hypothesis of homogeneity of proportional response are essentially the same—it is the perspective taken toward the research problem that is different. Second, if a Pearsonian χ^2 were applied to data in Table 3.4, a significant result would mean that observed proportions deviate from those expected, in the long run, under the hypothesis of homogeneity of proportions. Finally, unlike the earlier test of mutual independence, the present test on the hypothesis of homogeneity of proportions is clearly asymmetrycal. The sampling scheme used in the Goldsmid et al. study, for example, would not permit us to conclude that the proportion of distinguished teachers and the proportion of control teachers are homogeneous (or heterogeneous) over levels of the age variable.

Before we take leave of the Goldsmid et al. study, it may be of interest to note that the application of Eq. (3.8) to data presented in Table 3.3 results in a chi-square test statistic that is obviously nonsignificant, i.e., $\chi^2(2) = 1.31$, $p > .50$. The hypothesis that quality teaching is associated with faculty in mid-career finds little support in the data given in Table 3.5.

TABLE 3.5. Proportional Response to Levels of Age by Type of Teacher

Type of Teacher	Age Categories (B)			Marginals
	$B < 40$	$40 \leqslant B \leqslant 49$	$B > 49$	
Distinguished	.500 (.483)*	.333 (.308)	.167 (.208)	1.00
Control	.467 (.483)	.283 (.308)	.250 (.208)	1.00

*Expected cell proportions under the hypothesis of homogeneity of proportional response are contained in parentheses.

THE LIKELIHOOD-RATIO CHI-SQUARE

Many readers may be surprised to learn that there is an alternative to Pearson's goodness-of-fit χ^2 statistic. The alternative was developed by Pearson's pertinacious rival R. A. Fisher (1924) who, subsequent to developing the method of maximum likelihood that is described briefly in Chapter 2, derived a competing goodness-of-fit statistic based on the ML method. Despite differences in underlying method, however, Fisher's alternative, known as the maximum-likelihood-ratio chi-square, serves the same function as does the Pearsonian χ^2. The computational formula for the likelihood-ratio chi-square, symbolized by L^2 (although some writers prefer G^2), is

$$L^2 = 2\sum(f_{ij}) \left(\ln \frac{f_{ij}}{F_{ij}}\right) \tag{3.11}$$

As before, the summation is over all cells of the table. But unlike the calculation of the χ^2 statistic, the calculation of L^2 uses *natural* logarithms, abbreviated ln. Natural logarithms are logs to the base e, where $e = 2.718282$, and should not be mistaken for common logarithms, logs to the base 10.

Like the Pearsonian χ^2, L^2 is distributed approximately as a chi-square distribution when samples are sufficiently large. In fact, as samples become increasingly large, the χ^2 and L^2 converge or become asymptotically equivalent. In practice, therefore, both statistics will lead to essentially the same conclusions relative to goodness-of-fit. Because the calculation of L^2 has been perceived to be more involved than the calculation of χ^2, the former is not often seen in the research literature. However, L^2 possesses several properties that are most desirable in log-linear work, and hence it will soon prove to be our preferred goodness-of-fit statistic. For now, an illustration of the computation of L^2 on data shown in Table 3.4 will suffice.

Again, to subject to test the hypothesis that patterns of proportional response to the Age variable are homogeneous over faculty groups, the frequencies in Table 3.4 are entered into Eq. (3.11) as follows:

$$L^2 = 2 \left[(30) \left(\ln \frac{30}{29.0}\right) + (20) \left(\ln \frac{20}{18.5}\right) \right.$$
$$\left. + \cdots + (15) \left(\ln \frac{15}{12.5}\right) \right]$$

Performing the algebra yields

$$L^2 = 2[(30)(.0339) + (20)(.0780) + \cdots + (15)(.1823)]$$
$$= 2(.6596) = 1.32$$

This result compares well with the χ^2 value computed on the Goldsmid data.

RESIDUAL ANALYSIS AND OTHER FOLLOW-UP PROCEDURES

The statistical tests discussed so far are creatures of limited utility for, by themselves, they only enable us to reject the hypothesis under test. They do not, unfortunately, reveal to us either the direction or intensity of the relationship between two variables, or the boldness of a difference between two groups. For a test of mutual independence, a statistically significant χ^2 does not indicate the strength of the association between categorical variables or whether the association is significant in a practical sense. By the same token, a significant χ^2 associated with a test of homogeneity of proportions simply means that at least one group has demonstrated a response pattern that is different from the response pattern of at least one other group. The specific location and magnitude of difference remains unknown. Consequently, a thorough analysis of data usually requires some manner of follow-up.

The Analysis of Residuals

Simply put, residuals are differences between respective f_{ij}'s and F_{ij}'s. Therefore, it cannot be said that we are about to entertain an entirely new topic, for on several previous occasions we have looked at patterns of residuals to determine the directionality of an association that was found to be significant. Granted, these previous attempts at residual analysis were rather primitive for they amounted to scarcely more than a casual visual examination of patterns of raw residuals. In this section, more sophisticated methods of residual analysis will be introduced which can facilitate the interpretation of results gleaned from two-dimensional tables, especially tables containing numerous cells. But the benefits to be derived from a serious analysis of residuals increase in direct relation to the dimensionality of tables. Nevertheless, we choose the present simple context to introduce the basic properties and uses of two commonly used techniques for the analysis of residuals.

Standardized Residuals. In his authoritative paper, Haberman (1973) proposed a relatively simple definition of a residual, termed the standardized residual, which for the ijth cell of a two-dimensional table is

$$R_{ij} = (f_{ij} - F_{ij})/(F_{ij})^{1/2} \qquad (3.12)$$

To compute R_{ij} on the cell in the upper left corner in Table 3.3, for example, we get

$$R_{11} = (67 - 46)/\sqrt{46} = 3.10$$

The eight remaining R_{ij}'s may be calculated accordingly.

Let us make two observations about the contents of Eq. (3.12). The first is that a R_{ij} is a raw residual (i.e., $f_{ij} - F_{ij}$) that is divided by the square root of F_{ij} so as to effect an adjustment for the number of expected frequencies in the cell. Such an adjustment is intuitively pleasing. After all, to the extent to which an F_{ij} becomes large, a fixed raw residual becomes increasingly less indicative of a discrepancy between the sample result and the parameter posited by the null; therefore, the magnitude of the residual measure is adjusted downward. Of course, the converse results in an upward ajustment.

The second observation is that there appears to be a similarity in structure between the formula for standardized residuals and the formula for Pearson's χ^2 statistic. Compare Eqs. (3.3) and (3.12) and verify that, if the R_{ij}'s for all tabular cells are squared and then summed,

$$\chi^2 = \sum_{ij} (R_{ij})^2$$

But there is more to standardized residuals than this most interesting linkage to the χ^2. Haberman has shown that if the requisite conditions for the chi-square approximation are satisfied (conditions that were surveyed earlier in this chapter), then the probability distribution of R_{ij}'s tends toward the normal with a mean of zero and an asymptotic variance approaching, but not quite achieving, unity. In short, if requisites are met,

$$R_{ij} \sim N(0, \sigma^2 < 1)$$

Though not distributed precisely as a unit normal, if used judiciously the R_{ij}'s can function as standard normal deviates (i.e., z statistics) although they are apt to err in the direction of underestimating the nominal or "true" value of these deviates. Approximation to the standard normal is best when the numeric values of F_{ij}'s are relatively uniform and the number of levels comprising variables is large. All in all, standardized residuals possess great appeal. They are easily obtained with a pocket calculator and are even available on such computer programs as BMDP/4F and the MULTIQUAL program of Bock and Yates (1973). Further comment on their use in practice will follow the brief presentation on Freeman-Tukey deviates.

Freeman-Tukey Deviates. Prior to the introduction of standardized residuals, Freeman and Tukey (1950) showed that the quantity

$$\sqrt{f_{ij}} + \sqrt{f_{ij} + 1}$$

termed a *variance stabilizing transformation*, was distributed approximately as a normal distribution with a mean equal to $\sqrt{4F_{ij} + 1}$ and variance equal to 1. So by subtracting the mean of the probability distribution, an approximate z statistic is obtained. Accordingly, Freeman-Tukey deviates are given by

$$D_{ij} = \sqrt{f_{ij}} + \sqrt{f_{ij} + 1} - \sqrt{4F_{ij} + 1} \tag{3.13}$$

When we compute D_{ij} on the cell in the upper left corner of Table 3.3 we get

$$D_{11} = \sqrt{67} + \sqrt{67 + 1} - \sqrt{(4)(46) + 1}$$
$$= 2.83$$

It should be noted that there is less than perfect agreement between the R_{11} and D_{11} computed on the cell in question, i.e., $R_{11} = 3.10$, whereas $D_{11} = 2.83$. However, inasmuch as these two statistics only purport to approximate standard normal deviates, and only under optimal conditions, some manner of discrepancy between them is to be expected. An empirical comparison between standardized residuals and Freeman-Tukey deviates has been reported by Bishop et al. (1975). When both residual measures were computed on data in a four-dimensional table following the fit of a log-linear model, the two sets of deviates were almost identical. But, as those authors point out, the extremely large size of the sample ($n = 7,653$) likely contributed in large measure to the near perfect agreement between measures.

Let us emulate the work of Bishop et al. by comparing residuals on the much simpler data set provided by O'Connor and Sitkei (Table 3.3). R_{ij}'s and D_{ij}'s are juxtaposed in Table 3.6. It seems that, with the possible exceptions of cells $(AB)_{11}$ and $(AB)_{13}$, agreement between corresponding measures is quite good. Agreement is particularly good for those cells with high observed and expected frequencies. Further, in the aggregate a consideration of both measures leaves little doubt as to the location of major discrepancies between observed frequencies and those expected under the hypothesis of independence. Using a minimum residual distance of ± 2.00 as a rough indicator of statistical significance at the .05 level, significant departure from independence is seen in cells $(AB)_{11}$, $(AB)_{13}$, and $(AB)_{31}$.[4] Thus we can now conclude with even greater confidence that the O'Connor-Sitkei data indicate that it is in publicly supported institutions, not profit-making private institutions, where small numbers of retarded residents tend to be found.

It is hoped that the foregoing discussion and example have demonstrated the utility of an analysis of residuals for two-dimensional tables. As we shall see, residual analysis can be even more helpful with complex tables that are analyzed using log-linear models. Briefly, residual analysis can assist in the identification of outliers (highly deviant cell frequencies) due possibly to a clerical error or an anomaly in data. In either event, outliers can misrepresent the

---

[4] It is somewhat difficult and perhaps unwise to state an exact numerical criterion for the statistical significance of a residual because there are a number of mitigating factors to be considered that are difficult to quantify precisely. For example, since in a 3×3 table there are nine residuals, hence nine tests, good statistical practice would have us exercise some control over the Type I error rate which escalates with successive testing. But $\nu = 4$, and therefore in a sense there are but four tests that are independent; the remaining tests are somewhat dependent on the outcomes of the former. Even so, a conservative approach could be taken whereby the desired overall alpha risk (say $\alpha = .05$) is partitioned into four parts such that tests on residuals would be made at the .0125 level ($z_\alpha = \pm 2.24$) thus ensuring that the expected number of Type I errors for the four tests would not exceed .05. But on the other hand, both standardized and Freeman-Tukey deviates tend to be conservative approximations of standard normal deviates, reducing somewhat the need for explicit control over escalating alpha when the number of cells in a table is not too large. And since residuals should be viewed as suggestive indicators of departure, not definitive tests of departure, the compromise criterion of ± 2.00 seems reasonable here.

TABLE 3.6. Standardized Residuals and Freeman-Tukey Deviates as Applied to O'Connor-Sitkei Data in Table 3.3.

Type of Facility	Number of Residents (#R)					
	#R ≤ 10		10 < #R < 21		#R ≥ 21	
	R	D	R	D	R	D
Public	3.10	2.83	−.43	−.40	−2.96	−3.38
Nonprofit	−.10	−.08	−.68	−.67	.89	.99
Private Profit	−2.43	−2.59	1.17	1.16	1.35	1.32

adequacy of fit of a log-linear model. Detecting outliers and accounting for their presence may result in the acceptance of a model that fits a table reasonably well except for the deviant condition or conditions. A residual analysis can sometimes provide insights into reasons for the failure of a model to fit a table, insights that can point to a previously overlooked term which if incorporated into the model results in good fit (see Bishop et al., 1975, p. 138). In sum, by considering the analysis of residuals in this chapter, we have made an investment in our ability to undertake a log-linear analysis.

A Note on Other Follow-up Procedures

In addition to the study of residuals, if the mode of inquiry is symmetrical there are a number of procedures that attempt to measure the *strength of association* between variables in 2 × 2 tables. Perhaps most popular is the phi coefficient, a Pearson product-moment correlation coefficient computed on two dichotomous variables. Unfortunately, the phi coefficient is extremely sensitive to skewed marginal distributions, a limitation that phi shares with many other measures of association. To be more specific, the maximum possible numerical value of the phi coefficient is attenuated as marginal frequencies depart from an even split. An alternative measure of association for 2 × 2 tables is Yule's Q, as will be discussed at greater length in the next chapter. Yule's Q uses what is known as the *odds ratio* which is not adversely affected by uneven marginal distributions.

For the symmetrical analysis of tables larger than 2 × 2, less satisfactory measures of association have been proposed. Those most frequently used are Cramer's V and the contingency coefficient. Both measures are predicted on the χ^2 statistic and both range from zero (independence) to positive unity (perfect association). Both measures, however, are affected by unequal marginals in a manner similar to that of the phi coefficient. In addition, they are adversely affected when the contingency table is not square. For a detailed discussion of the merits and limitations of measures of association, the reader is referred to a monograph authored by Reynolds (1977). The brief texts by Maxwell (1961) and Everitt (1977) are also recommended.

Following-up the results of an asymmetrical analysis presents a different set of problems. But in recent years there have been advances in this area which in many ways make the follow-up process more tractable than it is for tests of association. For example, multiple comparison tests which are directly analogous to those used subsequent to an ANOVA (e.g., Tukey's test and the Scheffé) can be applied to groups to determine the specific nature of group differences with respect to their responses to a qualitative dependent variable. An extensive treatment of these post hoc multiple comparison procedures is provided by Marascuilo and McSweeney (1977, Chap. 6). Log-linear procedures, however, provide specific group effects (i.e., lambdas) thus minimizing the need to consult the procedures proposed by Marascuilo and McSweeney when a log-linear contingency analysis is performed. These latter procedures will be introduced in the next chapter.

4

Log-Linear Analysis
of Two-dimensional
Tables

The basic principles of log-linear analysis are introduced in this chapter. The importance of acquiring an understanding of these principles cannot be overstated, for as Knoke and Burke (1980, p. 7) have recently pointed out: "During the past decade a revolution in contingency table analysis has swept through the social sciences, casting aside most of the older forms for determining relationships among variables measured at discrete levels." The "revolution," of course, is the application of the log-linear models to the analysis of qualitative data, a revolution that has already occurred in sociology and that can be expected to gain significant momentum in other areas of behavioral inquiry during the 1980s.

It is fortunate that basic understandings of log-linear methods require little in the way of new learnings, for in a sense log-linear analysis may be viewed as the issue of a traditional marriage between the ANOVA and chi-square goodness-of-fit; or, from a less traditional perspective, it may be viewed as the product of the conventional ANOVA, the "newer" regression approach to ANOVA, and the chi-square goodness-of-fit. Hence, the intent of this chapter is to review salient features of these techniques and to show how these features are combined to provide the basis for log-linear methods.

The reader should be forewarned, however, that the revolutionary benefits associated with log-linear methods cannot be appreciated fully within the present context of two-dimensional contingency tables. In fact, if behavioral researchers were confronted *only* with two-dimensional tables in their work, there would be little or no need for the newer methods to be introduced here, for "the older forms" reviewed in the previous chapter would suffice. But because basic principles are best illustrated initially in simple contexts, we will begin our study of log-linear principles within the simpler context of two-dimensional tables. Keep in mind, however, that the contents of this chapter are designed as preparation for the use of log-linear models with contingency tables defined by more than two variables.

THE FOREST BEFORE THE TREES

The performance of a log-linear analysis is not a discrete act but rather is the engagement in a process that, like common factor analysis, is part art and part science. And to understand a process better, it often helps to first become acquainted with the major steps subsumed by the process. Therefore, let us outline in order the major operations that characterize log-linear analysis.

1. Define two or more qualitative variables and structure the constituent categories (or levels) of these variables so that they are mutually exclusive and exhaustive.

2. Determine the essential nature of the ensuing inquiry. That is, is the research seeking to identify relations between or among variables (i.e., symmetrical inquiry) or is the research attempting to identify differences between or among groups (i.e., asymmetrical inquiry)?

3. Specify the target and accessible populations. The target population refers to the population to which the researcher would like to be able to generalize his or her findings. The researcher often does not have complete access to the target population, however, and must therefore use a subset of the target population to construct the sampling frame. This subset of the target population, the population from which the researcher is able to draw a sample, is the accessible population.

4. Obtain a sample that is representative of the target population, if possible. If this is not possible, obtain a sample that is representative of the accessible population. For symmetrical inquiry, the overall sample size should be fixed and, if feasible, the sample should be drawn at random. If, however, the inquiry is asymmetrical, give serious consideration to fixing the size of respective levels of the

explanatory variable such that level sizes are equal or approximately equal.

5. Cross-classify members of the sample on the basis of the qualitative variables, being sure to observe the previously mentioned condition of response independence. Remember that a variable consisting of levels that are repeated measurements on the same subjects will likely violate the independence assumption and hence cast suspicion on the integrity of the analysis. Assuming that the condition of response independence is satisfied, resultant cross-tabulations or *observed* elementary cell frequencies are then arranged in a contingency table.

6. Specify a number of "ANOVA-like" models appropriate to the research so that a number of different sets of *expected* elementary cell frequencies can be generated for the table.

7. Compare the expected cell frequencies generated by each ANOVA-like model to the observed cell frequencies for agreement or goodness-of-fit.

8. Select from among the numerous models the model that is deemed *most acceptable*, i.e., the most parsimonious (or restricted) model that still represents an adequate fit.

9. Interpret the model deemed the most acceptable model in terms of either symmetrical concepts (e.g., mutual association, partial association) or asymmetrical concepts (e.g., main effects, interaction).

10. Employ follow-up procedures, if needed, either to estimate the strength of associations or to determine the specific nature of differences among groups.

The first five operations outlined above have been discussed in earlier chapters. The latter five operations are peculiar to log-linear analysis and constitute the subject matter of this chapter.

ANOVA-LIKE MODELS

Log-linear models resemble, in many important respects, the more familiar models associated with the ANOVA. A brief review of relevant ANOVA models is provided in this section. Readers who are not familiar with the models to be reviewed are encouraged to undertake independent study. Texts authored by Glass and Stanley (1970, Chap. 17), Hays (1973, Chap. 12), Kennedy (1978, Chaps. 4 and 6), and Winer (1971, Chap. 5), among others, contain extensive discussions of ANOVA models and related topics.

Fixed ANOVA Models for Cell Means

Consider a 2×2 table such as that given by Table 1.1, but assume momentarily that the four elementary cells of the table contain arithmetic *means* instead of observed cell frequencies. That is, until further notice, assume that we are working with interval (or ratio) data within the context of a 2×2 factorial design (with one measure per cell) instead of a fourfold contingency table. Moreover, assume that each of the four cells contains the *population* mean of measures within the four cells of the table. Since the cell means are parameters they will be denoted by μ_{ij}.

Now recall that in the traditional approach to the ANOVA, the μ_{ij}'s can be "explained" by a linear combination, or model, of the form

$$\mu_{ij} = \mu + \alpha_i + \beta_j + (\alpha\beta)_{ij} \tag{4.1}$$

which states that the population mean in the ith level of independent Variable A and the jth level of independent Variable B can be explained by the sum of four independent terms: the first is the parameter μ and the latter three are *effects*, i.e., differences between population means. More specifically, the anatomy of a μ_{ij} is given by the sum of:

1. The grand mean of the four population cell means, symbolized by μ. This parameter is a component of all cell means, and it serves as an anchor point, or point of departure, so that specific cell means can be defined in terms of subsequent effects or mean differences.

2. The effect associated with all population means belonging to the ith level of Variable A. This *main effect* is denoted by α_i, an abbreviation for $\mu_i - \mu$, the difference between the mean of all measures in the ith level of A and the grand mean. Recall that if the independent Variable A is fixed (as opposed to random), then $\Sigma\alpha_i = 0$.

3. The main effect associated with all population means belonging to the jth level of Variable B, symbolized by β_j, an abbreviation for $\mu_j - \mu$. Again, if B is fixed, $\Sigma\beta_j = 0$.

4. The effect associated with a particular cell, the ijth cell, due to the *interaction* of Variables A and B. The symbol for this effect is $(\alpha\beta)_{ij}$, and it is defined by the identity

$$(\alpha\beta)_{ij} = (\mu_{ij} - \mu) - (\mu_i - \mu) - (\mu_j - \mu)$$
$$= \mu_{ij} - \mu_i - \mu_j + \mu \tag{4.2}$$

If the interaction term is needed in the model it means that population cell means cannot be explained accurately by simply adding the effects of α_i and β_j. Note that interaction effects are *independent* of respective main effects. Therefore, to the extent to which inter-action between A and B is present, variability exists among the μ_{ij}'s subsequent to correcting for (or partialling out) the influences of the two main effects. Conversely, if A and B do not interact, then there will be no variability among the μ_{ij}'s subsequent to correcting for main effects. Further, if both independent variables are fixed, the interaction term is also fixed, and hence

$$\sum_i^a \sum_j^b (\alpha\beta)_{ij} = 0$$

Thus far, the μ_{ij}'s have been explained exclusively in terms of parameters. Yet, unless we are conducting a census, we do not have precise knowledge of the numerical values of the parameters. However, these parameters can be estimated from sample information. In the present situation, for example, the grand mean of a representative sample of observations would constitute the least-squares estimator of μ; the mean of all sample measures in the ith level of A is the least-squares estimator of μ_i, etc. That is, in terms of expectancy notation,

$$E(\bar{X}..) = \mu \qquad E(\bar{X}._j) = \mu_j$$
$$E(\bar{X}_i.) = \mu_i \qquad E(\bar{X}_{ij}) = \mu_{ij}$$

Therefore, by systematically substituting appropriate sample esti-mators for the parameters contained in Eq. (4.1), a *working model* may be obtained. The working model, which contains parameter estimators, is given by

$$\bar{X}_{ij} = \bar{X}.. + a_i + b_j + (ab)_{ij} \qquad (4.3)$$

where the sample estimate of the main effect for the ith level of A is a_i, an abbreviation for $(\bar{X}_i. - \bar{X}..)$; the estimate for the main effect of the jth level of B is b_j, short for $(\bar{X}._j - \bar{X}..)$; and the estimate of interaction is $(ab)_{ij}$, which in extended form is $(\bar{X}_{ij} - \bar{X}_i. - \bar{X}._j + \bar{X}..)$.

In the ANOVA, a model such as Eq. (4.1) is selected, the estimated effects specified by Eq. (4.3) are calculated, and these estimated effects are tested directly for statistical significance using

F tests. But unlike log-linear procedures, in the common practice of ANOVA the data analyst does not explicitly attempt to eliminate effects that are obviously nonsignificant to obtain a simpler, more parsimonious model. Such model simplification could be attempted, however. For example, if there were no interaction between *A* and *B*, the data analyst could advance a more economical model by eliminating the interaction term. Further, if interaction and the main effects for Variable *B* are nonexistent, then a model containing only μ and α_i to the right of the equals sign would suffice. Finally, should both main effects and the interaction be absent, cell means would be essentially equal and best explained by a model containing only μ on the right-hand side of the equation. Notice that each successive model posits an additional restriction; that is, each successive model sets an addition term or effect equal to zero. To the extent to which model effects are assumed to be zero and hence dropped from the model, the model is said to be *restrictive.*

Let us translate the models under discussion into *working models* (models containing parameter estimators instead of the parameters themselves) and unfold them in *hierarchical* fashion from the most restricted (hence, most parsimonious) to the least restricted (hence, least parsimonious):

$$\bar{X}_{ij} = \bar{X}.. \tag{4.4}$$

$$\bar{X}_{ij} = \bar{X}.. + a_i \tag{4.5}$$

$$\bar{X}_{ij} = \bar{X}.. + a_i + b_j \tag{4.6}$$

$$\bar{X}_{ij} = \bar{X}.. + a_i + b_j + (ab)_{ij} \tag{4.7}$$

The arrangement of models can also be said to be hierarchical because as we proceed from the most to the least restricted models, the more restricted models become proper subsets of the more complete models.

The important point is that the hierarchical arrangement of ANOVA working models is similar to the arrangement of models used in the analysis of two-way contingency tables. We will shortly modify the ANOVA working models so that they serve to explain or predict elementary cell *frequencies* instead of elementary cell *means.* Further, we shall see that models for contingency tables can take two forms: they can be structured *multiplicatively,* or they can be *additive.* In either case, they eventually produce identical expected cell frequencies (i.e., F_{ij}'s). Since models cast in the multiplicative are more compatible with the structure of hypotheses about qualitative data, we turn to these models first.

Multiplicative Cell Frequency Models

Unlike the ANOVA models of the previous section, models designed to explain elementary cell frequencies in two-way tables yield products, not sums. Aside from basic probability theory, most behavioral researchers are not used to working with multiplicative models and therefore are likely to be ill-at-ease upon first exposure to these strange creatures. Experience has shown, however, that with a little patience and practice, multiplicative models will not only become creatures of comfort but they will probably also become the intellectual vehicles most preferred by researchers when working with contingency tables.

The Mutual Equiprobability Model. Turning first to the cell-frequency model that parallels the ANOVA working model of Eq. (4.4), we have the single-parameter model

$$F_{ij} = \hat{\tau} \tag{4.8}$$

We interpret this model to mean that elementary cell frequencies are explained (or predicted) by only the parameter estimator $\hat{\tau}$. The tau parameter is the multiplicative counterpart to μ, the grand mean in ANOVA models. Specifically, the estimate of tau in Eq. (4.8) is the grand *geometric* mean of expected cell frequencies given by the model in question.

Before proceeding, we pause to illustrate and define the geometric mean. Suppose that we had four scores (i.e., X_i's) with values of 33, 7, 37, and 23. Since $n = 4$ here, the arithmetic mean is: $\bar{X} = \Sigma X_i/n = 100/4 = 25$. The geometric mean, however, is the nth root of the *product* of the four scores. That is, in general,

$$\bar{G} = \left(\prod_{i}^{n} X_i \right)^{1/n} \tag{4.9}$$

where $\bar{G}$ is used as a convenient symbol to denote the geometric mean. Thus, for the specific values in question,

$$\bar{G} = \left(\prod_{i=1}^{4} X_i \right)^{1/4} = (33 \cdot 7 \cdot 37 \cdot 23)^{1/4} = (196,581)^{1/4}$$

$$= 21.0565$$

One can see why the geometric mean can be said to be the multiplicative counterpart of the arithmetic mean.

One difference, therefore, between ANOVA models and multiplicative contingency table models is that the anchor point (i.e., the first parameter estimator) of the latter is a geometric mean, not an arithmetic mean. Other differences exist, however. Perhaps the most striking difference is that in the ANOVA, one model is posited and *all* (or most) observed data are used to make a simultaneous judgement about all terms in the model. In marked contrast, in log-linear work multiple models are posited. Moreover, only one of these models (the saturated model) uses all observed data upon which to base model estimates. Model (4.8), for example, uses only one piece of observed information, the size of the sample, to determine the numerical value of its single-parameter estimate. Thus, for the data given in the fourfold table of Chapter 1 (see Table 1.1), Model (4.8) uses only the fact that $n = 100$ to estimate τ, the common *expected* frequency for the four elementary cells. As we shall see, as contingency table models become less restricted, a correspondingly greater amount of observed information is used to estimate model parameters.

Still another important difference between ANOVA and contingency table models is found in the way that these models estimate parameters. Essentially, least-squares procedures underlie estimation in the ANOVA, whereas maximum-likelihood (ML) estimation is generally used for contingency tables.

To appreciate this distinction and the manner in which the F_{ij}'s are generated by Model (4.8), consider the following simple exercise. Suppose that you were asked to estimate the numerical values of four scores but all you knew about the scores was that they sum to 100. Most likely you would estimate each of the scores to be 25, the arithmetic mean of the scores. This also happens to be the least-squares estimate, for in the long run the sum of squared differences between actual values and estimated values will be a minima. Of course, when dealing with scores within an ANOVA situation, least-squares estimation is appropriate. But consider a situation where data are qualitative. Suppose that we had a fourfold table and we know only that $n = 100$, and we want to estimate how many cases to expect in each of the four cells. Because we are essentially dealing with proportions, ML estimation is most appropriate here. Further, from our reading of Chapter 2, we know that the ML estimate of the frequency that can be expected to belong to the cell in the ith level of A and jth level of B would be

$$F_{ij} = n/ab \tag{4.10}$$

which, for the current exercise, is

$$F_{ij} = 100/4 = 25$$

In sum, we would deem that outcomes most likely in the long run are best obtained by distributing the n cases uniformly among the ab cells such that, in this case, each cell contains 25 subjects.

Essentially we have just done the work of Model (4.8). Using only one observed datum, total sample size, we have used ML procedures to generate four expected cell frequencies. Having generated the four F_{ij}'s, the tau parameter in Model (4.8) can be more readily defined. Specifically, the tau parameter is the geometric mean of expected elementary cell frequencies advanced by the model using maximum-likelihood estimation procedures. In more general terms, expected cell frequencies provided by Model (4.8) are given by

$$F_{ij} = \hat{\tau} = \bar{G} = \prod_{ij}^{ab} \left(\sum_{i}^{a} \sum_{j}^{b} f_{ij}/ab \right)^{1/ab}$$

$$= [(n/ab)^{ab}]^{1/ab}$$

$$= n/ab \tag{4.11}$$

Recapping our work to this point, the application of Model (4.8) to the data in Table 1.1 gives

$$F_{ij} = \prod_{ij}^{4} (100/4)^{1/4} = [(100/4)^{4}]^{1/4}$$

$$= 100/4 = 25$$

a solution that we now recognize as the geometric mean of expected cell frequencies given by the model in question.

The question that arises at this point is how well the F_{ij}'s produced by Model (4.8) fit the observed f_{ij}'s reported in Table 1.1. We will eventually perform a goodness-of-fit test between these expected and observed frequencies, and should it turn out that the fit is "good" (which is not going to be the case), then Model (4.8) will be adopted as the best explanation for observed data.

If Model (4.8) were to be adopted, what kinds of conclusions could be advanced? From a symmetrical perspective, it is clear that we could conclude that there is no association between Variables A (Sex) and B (Attitude). Variables A and B are not only independent, but since there are 25 fitted cases in each cell, this outcome can be labeled one of *mutual equiprobability*, meaning that every sample subject has an equal probability of being assigned to any of the four cells in the table—a probability of .25 in this case. From an asymmetrical perspective, with Variable A the explanatory variable, not only are there no differences between females and males relative to their proportional response to respective categories of Variable B, but we can also say that their responses are equiprobable (i.e., distributed evenly) over categories of the response variable. Thus, should this model be selected for interpretation, it follows that we would either conclude that Variables A and B are strikingly independent, or that there are no differences between females and males with respect to their attitudes concerning the antiabortion amendment. For these reasons, single-parameter models will be called either models of equiprobability (symmetrical) or completely null models (asymmetrical).

The Conditional Equiprobability Model. Consider next the multiplicative version of Eq. (4.5), which is

$$F_{ij} = \hat{\tau}\hat{\tau}_i^a \tag{4.12}$$

Before defining the two factors in this model, let us establish three requisite understandings.

First, whereas completely null models use only n to fit F_{ij}'s, this model uses [A], the main marginal frequencies observed for Variable A, to fit F_{ij}'s. (Note that in Table 1.1, [A] consists of $f_1^a = 40$ and $f_2^a = 60$.) The second understanding is that when the frequencies in [A] are transformed into proportions ($p_1^a = .40$ and $p_2^a = .60$), the sample proportions are ML estimates of respective population proportions and they follow the multinomial law. Finally, the use of p_i^a's as estimators of P_i^a's is the only constraint that Model (4.12) imposes on data. Therefore, given that $p_1^a = .40$ and $p_2^a = .60$, maximum-likelihood thinking holds that *all remaining cell entries are equiprobable*. Hence, the fitted F_{ij}'s given by the model in question are:

	B_1	B_2	$[A]$
A_1	20	20	40
A_2	30	30	60

We now define the estimators appearing in Eq. (4.12). The first parameter on the right represents the geometric mean of cell frequencies fitted by the model. More precisely,

$$\hat{\tau} = \bar{G} = \left(\prod_{ij} F_{ij} \right)^{1/ab} = (20 \cdot 20 \cdot 30 \cdot 30)^{1/4} = (360,000)^{1/4}$$

$$= 24.4951$$

The second parameter acknowledges main-marginal differences in $[A]$ and therefore embraces two specific estimates, namely τ_1^a and τ_2^a. The estimate associated with the ith category of Variable A may be defined by

$$\hat{\tau}_i^a = \bar{G}_i^a / \bar{G}$$

where $\bar{G}_i^a$ is the geometric mean of all fitted cell frequencies appearing in the ith level of A. Thus, $\hat{\tau}_1^a$ is

$$\hat{\tau}_1^a = \bar{G}_1^a / \bar{G} = \left(\prod_i^a F_{1j} \right)^{1/a} / \bar{G}$$

$$= (20 \cdot 20)^{1/2} / 24.4951 = 20 / 24.4951$$

$$= .8165$$

The second specific parameter estimate is

$$\hat{\tau}_2^a = \bar{G}_2^a / \bar{G} = (30 \cdot 30)^{1/2} / 24.4951$$

$$= 1.2247$$

It was in fact not necessary to calculate directly the second specific parameter, because, as will become apparent, the product of specific parameters in multiplicative contingency table models equals unity. That is,

$$\prod_i^a \hat{\tau}_i^a = 1.00 \qquad (4.13)$$

Exploiting this fact, $\hat{\tau}_2^a$ is obtained most easily by taking the reciprocal of $\hat{\tau}_1^a$, as indicated below:

$$\hat{\tau}_2^a = 1/\hat{\tau}_1^a = 1/.8165 = 1.2247$$

Pause to note that

$$\prod_i^a \hat{\tau}_i^a = (.8165)(1.2247) = 1.00$$

Some readers may be experiencing difficulty in grasping the underlying meaning of the factors contained in the model. However, when we convert these multiplicative models to linear models, these factors will be shown to be directly analogous in meaning to their more familiar counterparts in ANOVA models.

Since we have determined the numerical values of parameter estimates for Model (4.12) as they apply to data in Table 1.1, as an exercise let us use this acquired knowledge to generate at least one expected cell frequency, say F_{11}. Substituting known values into the equation for Model (4.12) gives

$$F_{11} = \hat{\tau}\hat{\tau}_1^a = (24.4951)(.8165) = 20.0002$$

where a slight discrepancy is seen due to rounding error. Remaining frequencies fitted by this model can be generated by similar methods, and, as expected, they turn out to be those given in the unnumbered table presented several paragraphs earlier.

Eventually, we will want to determine if the F_{ij}'s fitted by Model (4.12) constitute an "acceptable" fit. If they do (but they will not) then Model (4.12) will be adopted and interpreted. And should this model be adopted, the omnibus conclusion that will be advanced will be one of *conditional equiprobability* for the symmetrical case. That is, if an adjustment is made for the inequalities in the marginals for Variable A, then both females and males in the sample have an equal probability—a probability of .50—of being placed in either the first or second category of Variable B. To illustrate, suppose we adjust for unequal sexual representation by regarding each sexual group as an intact entity and examine not frequencies but rather proportional response *within* levels of A. To do this, set $p_1^a = 1.00$ and $p_2^a = 1.00$. Transforming the fitted frequencies *within* levels of A to proportions would result in $\hat{P}_{11} = .50$, $\hat{P}_{12} = .50$, $\hat{P}_{21} = .50$, and $\hat{P}_{22} = .50$. From a symmetrical perspective, the outcome indicated above represents a special case of independence between Variables A and B. By the same token, if inquiry is asymmetrical, it is clear that females and males do not differ in response to Variable B; hence, the null hypothesis of no main effects due to sex should be retained.

The Mutual Independence Model. Consider next a model that fits both $[A]$ and $[B]$, a model that is structurally analogous to the ANOVA Model (4.6).

$$F_{ij} = \hat{\tau}\hat{\tau}_i^a\hat{\tau}_j^b \tag{4.14}$$

Looking ahead, should observed data in Table 1.1 turn out to be most consistent with this model, the two principal hypotheses tested by traditional chi-square procedures would be retained. That is, Variables A and B would be regarded as being independent or the hypothesis of homogeneity of response will be retained.

To see why this is so, we use Model (4.14) to generate F_{ij}'s for our working example. As noted, this model employs only knowledge of $[A]$ and $[B]$. In the example, $[A]$ contains $f_1^a = 40$, $f_2^a = 60$ and $[B]$ is $f_1^b = 70$, $f_2^b = 30$. Thus, with knowledge of only the observed marginals, marginals that can be readily converted into proportions and hence serve as ML estimates of respective population proportions, the probability that a member of the sample will be jointly classified in the ijth cell is

$$\hat{P}_{ij} = p_i^a p_j^b$$

which we recognize as a restatement of Eq. (3.6) from Chapter 3. Recall that Eq. (3.6) is the mathematical definition of response independence in two-way tables. Clearly, independence is inherently a multiplicative concept, and hence an argument in favor of working with models that are multiplicative in nature. For data at hand, F_{ij}'s are produced by Eq. (3.6) or its equivalent, Eq. (3.7), with the following result:

	B_1	B_2	$[A]$
A_1	28	12	40
A_2	42	18	60
$[B]$	70	30	100

If the reader has any doubt as to how these values were obtained, a review of the discussions in Chapter 3 associated with Eqs. (3.5), (3.6), and (3.7) is recommended.

Having used Model (4.14) to fit a set of F_{ij}'s, let us examine more closely the parameter estimates in the model. The first factor to the right of the equals sign is $\hat{\tau}$, the geometric mean of F_{ij}'s produced by the model. It is

$$\hat{\tau} = \bar{G} = (28 \cdot 12 \cdot 42 \cdot 18)^{1/4} = 22.4502$$

The estimate of the *marginal effect* associated with the first level of Variable A is

$$\hat{\tau}_1^a = \bar{G}_1^a/\bar{G} = (28 \cdot 12)^{1/2}/22.4502 = 18.3303/22.4502$$
$$= .8165$$

Hence, the marginal effect for the second level of A is

$$\hat{\tau}_2^a = \bar{G}_2^a/\bar{G} = 1/\hat{\tau}_1^a = 1/.8165$$
$$= 1.2247$$

The marginal effects for Variable B are obtained by parallel calculations:

$$\hat{\tau}_1^b = \bar{G}_1^b/\bar{G} = (28.42)^{1/2}/22.4502 = 1.5275$$
$$\hat{\tau}_2^b = \bar{G}_2^b/\bar{G} = 1/\hat{\tau}_1^b = .6547$$

Note that the product of the effects for both A and B equals unity.

Before considering the next model, let us put into perspective our work to this point. The three models presented to this point were relatively simple and therefore we have been able to use elementary probability theory to advance the F_{ij}'s associated with each. Irrespective of the method used to obtain F_{ij}'s, once they are obtained the effect parameters in the model can be determined by the formulas that we have been using. Moreover, we should be able to use the effect parameters to fit a frequency in any cell of the table. For example, the F_{21} given by Model (4.14) is

$$F_{21} = \hat{\tau}\hat{\tau}_2^a\hat{\tau}_1^b = (22.4502)(1.2248)(1.5275)$$
$$= 42.0017$$

where it appears that we have an accumulated error of .0017 due to rounding. Finally, realize that the F_{ij}'s given by Eq. (4.14) will be compared eventually to the f_{ij}'s contained in Table 1.1, and should the F_{ij}'s constitute an acceptable fit, one of two omnibus deductions will be made: either (a) Variables A and B will be considered mutually independent or (b) there is no evidence of a significant main effect in proportional response to Variable B due to Variable A.

The Saturated Model. The multiplicative analogue to the complete ANOVA model, Model (4.7), is termed the saturated model. This model is written as

$$F_{ij} = \hat{\tau}\hat{\tau}_i^a\hat{\tau}_j^b\,\hat{\tau}_{ij}^{ab} \qquad (4.15)$$

Like its ANOVA counterpart, but unlike previous multiplicative models, the saturated model makes use of all observed tabular data. Therefore, the elementary cell frequencies produced by this model are identical to f_{ij}'s; that is, for this model, $F_{ij} = f_{ij}$ for all values of i and j, producing a perfect fit. Since neither theory nor mathematics are needed to generate F_{ij}'s, we proceed immediately to an explication of model parameters.

As expected, the first factor, $\hat{\tau}$, is the geometric mean of f_{ij}'s in Table 1.1:

$$\hat{\tau} = \bar{G} = (33 \cdot 7 \cdot 37 \cdot 23)^{1/4} = 21.0563$$

The marginal effects for Variable A are

$$\hat{\tau}_1^a = \bar{G}_1^a/\bar{G} = (33 \cdot 7)^{1/2}/21.0563 = .7218$$
$$\hat{\tau}_2^a = 1.3854$$

Similarly, effects for B are

$$\hat{\tau}_1^b = \bar{G}_1^b/\bar{G} = (33 \cdot 37)^{1/2}/21.0563 = 1.6595$$
$$\hat{\tau}_2^b = .6026$$

This brings us to the third and new factor in the model, a factor which corresponds to first-order interaction in conventional ANOVA models. The synonymity between first-order effects in ANOVA and in the log-linear will be explored in detail in the next section. For now, we simply point out that interaction effects for cells of a contingency table may be estimated by

$$\hat{\tau}_{ij}^{ab} = f_{ij}/(\hat{\tau}\hat{\tau}_i^a\hat{\tau}_j^b) \tag{4.16}$$

To illustrate, we calculate the interaction effect for the cell in the first level of Variable A and first level of B.

$$\hat{\tau}_{11}^{ab} = 33/[(21.0563)(.7218)(1.6595)]$$
$$= 33/25.2218 = 1.3084$$

The remaining three effects need not be calculated directly for, as in the case of a 2×2 ANOVA, once the value of one effect is known, the remaining three are determined. They are determined because the marginal products of multiplicative interaction effects turn out to be unity. Stated mathematically, for all values of i,

$$\prod_j^b \hat{\tau}_{ij}^{ab} = 1.00$$

and for all values of j,

$$\prod_i^a \hat{\tau}_{ij}^{ab} = 1.00$$

Consequently, for fourfold tables, it follows that

$$\hat{\tau}_{11}^{ab} = \hat{\tau}_{22}^{ab}$$

and

$$\hat{\tau}_{12}^{ab} = \hat{\tau}_{21}^{ab} = 1/\hat{\tau}_{11}^{ab} = 1/\hat{\tau}_{22}^{ab}$$

Hence, for the working example, $\hat{\tau}_{11}^{ab} = \hat{\tau}_{22}^{ab} = 1.3084$ and $\hat{\tau}_{12}^{ab} = \hat{\tau}_{21}^{ab} = .7643$. A summary of our work with these interaction effects is displayed below:

	B_1	B_2	Marginal Products
A_1	1.3084	.7643	1.0000
A_2	.7643	1.3084	1.0000
Marginal Products	1.0000	1.0000	1.0000

Finally, since all parameter estimates for the saturated model have been calculated, as an exercise the reader is encouraged to use the model to produce the four expected cell frequencies. If done correctly, the F_{ij}'s will be equivalent to the f_{ij}'s appearing in Table 1.1.

A few words should be said about the consequences of selecting the saturated model as the best representation of resultant data. Since interaction effects are, in effect, deviations from mutual independence, adoption of this model indicates either rejection of the independence hypothesis or rejection of the hypothesis of homogeneity of proportional response. Thus, if the inquiry is symmetrical, acceptance of Model (4.15) means that Variables A and B are associated; if the inquiry is asymmetrical, the selection of the saturated model means that main effects exist for levels of the explanatory variable—females and males do not respond similarly on the response variable. In a comparative sense, the traditional chi-square procedures of Chapter 3 represent a special case of the more comprehensive methodology introduced here. When performing a traditional chi-square analysis, implicitly either Model (4.14) or Model (4.15) is selected and then interpreted. Log-linear methods make model selection and interpretation explicit.

Log-Linear Cell Frequency Models

Because multiplicative models are true to the concepts underlying explanations for contingency table data, in this book they have been chosen as the first to be discussed. Linear models, however, can also be written to explain cell frequencies in contingency tables. Linear models are not distinctive, they are simply multiplicative models that become linear in their logarithms. Linearity is a ubiquity in the behavioral sciences. It is not surprising, therefore, that most behavioral researchers prefer to think and work with linear phenomena. Consequently, most of our statistical techniques and most of our computer software is based on linear thinking and linear solutions.

Thus, despite this writer's penchant for the multiplicative models of the last section, it must be conceded that by casting our models in the form of linear combinations, a number of advantages will be accrued.

To transform a multiplicative model into the linear, the numerical value of each constituent factor is expressed in terms of its natural logarithm (ln), a logarithm to the base e. (Hence, the term "log-linear" as the descriptor of these models.) Accordingly, instead of multiplying parameter estimates to obtain a product, log estimates will be placed in linear combination yielding a sum. Also, the operations of raising a factor to a specified power or exponent, or of extracting from a factor a specified root, will be replaced by the corresponding operations of multiplication and division.

By way of illustration, consider the single parameter model presented as Eq. (4.8). Upon transformation, the log-linear version is

$$\ln F_{ij} = \hat{\lambda} \tag{4.17}$$

The right-hand term is the natural log of $\hat{\tau}$, i.e., $\hat{\lambda} = \ln \hat{\tau}$. It follows that cell frequencies fitted by this model will also be natural logs. (This means that actual cell frequencies will be respective antilogarithms, numbers corresponding to given logarithms.) Hence, for the example where the $\hat{\tau}$ of Model (4.8) was found to be equal to 25, the equivalent log-linear model is

$$\ln F_{ij} = \ln (25) = 3.2189$$

This model fits each cell with a value of 3.2189, the antilog of which is 25, and thus exhibits a condition that we earlier called mutual equiprobability.

If the three remaining multiplicative models—Models (4.12), (4.14), and (4.15)—are each in turn subjected to a logarithmic transformation, the following results are obtained:

$$\ln F_{ij} = \hat{\lambda} + \hat{\lambda}_i^a \tag{4.18}$$

$$\ln F_{ij} = \hat{\lambda} + \hat{\lambda}_i^a + \hat{\lambda}_j^b \tag{4.19}$$

$$\ln F_{ij} = \hat{\lambda} + \hat{\lambda}_i^a + \hat{\lambda}_j^b + \hat{\lambda}_{ij}^{ab} \tag{4.20}$$

The saturated model in log-linear form provides us with an opportunity to explore further the meaning of model parameters and effects. Recall that in the multiplicative, the following expression was used to obtain $\hat{\tau}_i^a$, the estimated effect of the ith level of Variable A:

$$\hat{\tau}_i^a = \bar{G}_i^a / \bar{G}$$

Effecting a logarithmic transformation on the above results in

$$\hat{\lambda}_i^a = \ln \hat{\tau}_i^a = \ln \bar{G}_i^a - \ln \bar{G}$$

Now, recall that in the ANOVA the corresponding effect is

$$\hat{\alpha}_i = \hat{\mu}_i - \hat{\mu}$$

Hence we may infer (and correctly so) that the estimator $\hat{\lambda}_i^a$ is directly analogous to its corresponding member in conventional ANOVA models.

Let us see if this direct correspondence also holds for interaction effects. In the multiplicative an unspecified interaction effect was given by Eq. (4.16),

$$\hat{\tau}_{ij}^{ab} = f_{ij} / (\hat{\tau} \hat{\tau}_i^a \hat{\tau}_j^b)$$

But in the log-linear, an unspecified interaction effect is

$$\begin{aligned} \hat{\lambda}_{ij}^{ab} &= \ln \hat{\tau}_{ij}^{ab} = \ln f_{ij} - (\ln \hat{\tau} + \ln \hat{\tau}_i^a + \ln \hat{\tau}_j^b) \\ &= \ln f_{ij} - \ln \bar{G} - \hat{\lambda}_i^a - \hat{\lambda}_j^b \\ &= \ln f_{ij} - \ln \bar{G}_i^a - \ln \bar{G}_j^b + \ln \bar{G} \end{aligned} \qquad (4.21)$$

Now, recall that in the ANOVA the interaction effect linked to the ijth cell may be defined as the difference between μ_{ij} and μ, subsequent to partialling out appropriate main effects. That is, first-order interaction in the ANOVA is commonly defined as

$$\begin{aligned} (\hat{\alpha}\hat{\beta})_{ij} &= \hat{\mu}_{ij} - \hat{\mu} - \hat{\alpha}_i - \hat{\beta}_j \\ &= \hat{\mu}_{ij} - \hat{\mu}_i - \hat{\mu}_j + \hat{\mu} \end{aligned}$$

Clearly, a comparison of the lines above with the lines of Eq. (4.21) shows that *effect parameters in log-linear models may be legitimately viewed and subsequently interpreted in a manner consistent with conventional ANOVA models.* Although alternative views and interpretations exist (e.g., an odds ratio interpretation), and these alternative views have their advocates (cf. Goodman, 1972b), it is my considered belief that consciously exploiting the many synonymous features of the ANOVA and the log-linear has great merit, particularly for researchers in areas where historically the ANOVA has been a most comfortable and reliable workhorse.

Wrapping Up Some Loose Ends

It was said that the first major operation peculiar to log-linear analysis was the specification of ANOVA-like contingency table models. In this section, we surveyed models for two-dimensional tables. We began with a model that contained only one parameter estimator and added estimators in ascending order so that each ensuing model contained *all* the parameter estimates of the previous models plus the newly incorporated term (or factor). Simply put, the models were addressed hierarchically, from most to least restrictive. Normally, in the absence of strong a priori theory, the specification of models hierarchically is most desirable, for this manner of specification minimizes the chances of overlooking important effects and maximizes the ease of computing expected cell frequencies.

A summary of hierarchical models for two-dimensional tables is contained in Table 4.1. Space limitations discourage a presentation in this table of models that are linear in their logarithms. Relative to the table, be sure to notice that for future reference the models of Table 4.1 have been renumbered (1) through (4), a numbering system that conveniently reflects the number of *general* (not basic) parameters in the models.[1] Note too that carets (ˆ) no longer appear over terms in the models because it is now well understood that all model terms are estimators of parameters and not parameters per se. Finally, observe that verbal descriptions of hypotheses associated with each model for both a symmetrical and an asymmetrical analysis are contained in Table (4.1). These descriptions will assume greater importance in ensuing discussions.

Before entertaining the next operation, that of comparing models for goodness-of-fit, a number of important differences between conventional ANOVA models and contingency table models

---

[1] The number of general parameters is equivalent to the number of terms (or factors) in the model. On the other hand, the number of basic parameters refers to the number of estimates and effects that are free to vary. Relative to a 3×2 table, for example, there are four general parameters but six basic parameters in the saturated model. This distinction will be discussed in the next chapter.

TABLE 4.1. Hierarchical Multiplicative Models for Two-dimensional Tables

Model Number	Models	Fitted Marginals	Associated Omnibus Hypotheses	
			Symmetrical	Asymmetrical
(1)	$F_{ij} = \tau$	n	Equiprobability	
(2)	$F_{ij} = \tau\tau_i^a$	$[A]$	Conditional equiprobability	
(3)	$F_{ij} = \tau\tau_i^a\tau_j^b$	$[A], [B]$	Mutual independence	Null
(4)	$F_{ij} = \tau\tau_i^a\tau_j^b\tau_{ij}^{ab}$	$[AB]$	Nonindependence	Main effects for A

Note. For the asymmetrical case, it is assumed that Variable B is the response or logit variable. For this case, Models (1) and (2) are not legitimate logit models.

should be made more explicit. An obvious difference is that ANOVA models represent explanations of cell means for a dependent variable that can be measured on either an interval or a ratio scale, whereas contingency table models describe cell counts or frequencies in terms of the relations among variables. Another obvious difference is that ANOVA models use all observed data to estimate parameter effects, whereas contingency table models use progressively more observed data as terms are added to the models. Only the saturated model, Model (4) in Table 4.1, makes use of all data by fitting elementary cells in the [AB] configuration.

A central aim in most statistical work is to assess (or test) model terms for their importance (or statistical significance). In the ANOVA, for example, a model is adopted and during the analysis an estimate of error variability (i.e., mean square error) is obtained which makes it possible to test directly each parameter estimate for statistical significance. Testing takes place *within* the system of analysis prescribed by the model. Estimates that do not achieve significance are excluded from the model only in the sense that they are generally ignored in subsequent substantive conclusions. A much different strategy, however, is employed in log-linear work. A different strategy is required because log-linear models do not have built-in error terms; therefore, it is not possible to obtain estimates of error variability within the structure of the models per se. To overcome this apparent liability, a series of models is specified, each model containing a different set of parameter estimates. A judgement as to which estimates are important is made by determining which model in the series gives F_{ij}'s that fit observed data reasonably well. Hence, subsequent to model specification, the major task becomes one of examining each model for "goodness-of-fit" and selecting the model that appears to be "most acceptable." Since all models other than the saturated one posit at least one distinctive null hypothesis, and since a nonsaturated model often is selected as being most acceptable, it follows that a null hypothesis often will be accepted as the most reasonable explanation of factors affecting observed data.

This model-fitting strategy, which often results in the "acceptance of a null hypothesis," presents intellectual problems for many researchers, for many of use have been taught to advance only those findings that emanate from a clear *rejection* of a null hypothesis. Classical hypothesis testing, to be sure, has merit; but exclusive reliance upon traditional approaches is not always good science. Good science is the intelligent pursuit of explanations for phenomana —it is not blind adherence to prescribed method. The model-fitting approach to be introduced next, and the process of selecting models to be introduced shortly, together represent an attempt to pursue intelligent explanation. It is hoped that following these exposures, reservations currently held by some readers will be assuaged.

COMPARING MODELS FOR GOODNESS-OF-FIT

Since the four models in Table 4.1 give different expected elementary cell frequencies, traditional chi-square procedures, reviewed in Chapter 3, can be used to determine the extent to which expected frequencies correspond to observed frequencies. In our work, chi-square procedures will be applied in two different ways. The first is a straightforward application of either Pearson's χ^2 statistic (Eq. 3.3) or the likelihood-ratio L^2 statistic (Eq. 3.11), or both, to test for overall agreement between model F_{ij}'s and observed f_{ij}'s. This first use of the chi-square is often simply referred to as "goodness-of-fit," although some writers refer to this application as *residual* chi-square testing. Since the term "residual" lends itself to a more restrictive meaning, it will be used throughout this text to denote chi-square tests of overall agreement.

The second use of the chi-square is specific to the L^2 statistic and generally involves comparisons of models in pairs. Because the likelihood-ratio L^2 possesses additive properties, the arithmetic difference between the residual L^2 statistics of two models can be tested for statistical significance to determine if the more parsimonious member of the pair fits data significantly less well then the less parsimonious member of the pair. Differences between residual L^2's are termed *component* chi-square values, and they will prove to be most helpful in narrowing the range of models that are potentially acceptable.

Residual Chi-square Comparisons

As mentioned, residual chi-squares are used to assess the extent of agreement between cell frequencies that are actually observed and those issued by a model or models. To illustrate, the hypothetical data in Table 1.1 will again be used. Accordingly, the expected cell frequencies generated by each of the four models have been determined. Recall that Model (1), the completely null model, gave F_{ij}'s equal to 25 for all four cells; the results of Model (2) were $F_{11} = F_{12} = 20$ and $F_{21} = F_{22} = 30$; and Model (3) yielded $F_{11} = 28$, $F_{12} = 12$, $F_{21} = 42$, and $F_{22} = 18$. Of course, the F_{ij}'s of the saturated model, Model (4), perfectly reproduced the observed cell frequencies of Table 1.1. To make residual comparisons, we invoke Eqs. (3.3) and (3.11) to calculate the magnitudes of Pearsonian and likelihood-ratio statistics, respectively. The results of these calculations for each model in Table 4.1 are summarized below:

	Residual Values	
Model	χ^2	L^2
(1)	21.44	25.68
(2)	20.17	21.65
(3)	4.96	5.19
(4)	0.00	0.00

As evidenced by its residual chi-square, the fit provided by the saturated model is perfect, i.e., $L_4^2 = 0.00$. This does not necessarily mean, however, that this model will be chosen as "most acceptable." Granted, observed data are completely captured by saturated models, but observed data (and saturated models) generally are redolent with complexity. Our goal then, and that of scientific inquiry in general, will be to try to define and describe attributes, measured by data, in the fewest number of terms, in the simplest of structures, without distorting or sacrificing meaning. Our objective is to seek parsimony of explanation, hence we strive to choose models that contain only parameters essential to functional explanation. Thus, upon assessment of log-linear models, it might turn out that a model containing fewer terms (or factors) than the saturated may fit observed data almost as well. If this should be the case, because of its simplicity, it would be considered more acceptable. Looking at the three unsaturated models reveals that, as expected, the fit becomes increasingly less good as models contain fewer terms. Understandably, the most restricted model, the completely null model, shows the poorest fit, $L_1^2 = 25.68$.

Formal tests of statistical significance often assist the researcher in making decisions. To statistically test for agreement between f_{ij}'s and the various sets of F_{ij}'s, the proper number of degrees of freedom must be determined for each residual chi-square test. A number of "rules of thumb" have been proposed to assist in this determination (cf. Goodman, 1970, p. 231; Fienberg, 1977, p. 36). Before we propose a most simple set of rules to determine the number of degrees of freedom, let us employ simple logic to discern them for residual tests in fourfold tables.

We begin by viewing the cell frequencies of the 2×2 table as a sample of observations of size four. For each model under test, we will determine the number of cell frequencies that are free to assume any numerical value, given the constraints imposed by the model in question. This number is equivalent to ν, the appropriate number of degrees of freedom for the model in question. For example, if there were no model, there would be no constraints and cell frequencies would be free to assume any value. It follows, therefore, that $\nu_0 = 4$ in this suppositional situation. Not suppositional is Model (1) where F_{ij}'s have been fitted using knowledge of n. In theory, three of the cell frequencies are free to vary here; the fourth is not, for it must assume a specific value to make the sum of frequencies equal to n. Hence, the residual chi-square test for Model (1) will be made using $\nu_1 = 3$. With respect to Model (2), since $[A]$ is used to fit F_{ij}'s, cell frequencies *within* each level of Variable A must sum to a f_i^a. To comply with this, only one cell frequency within each level of A can be arbitrarily specified; consequently, $\nu_2 = 2$. Model (3) fits both the observed main-marginals of Variables A and B, and as a result only one cell frequency can be arbitrarily assigned a value, for once this is done, the remaining three frequencies will have determined values under the constraint that levels of A sum to $[A]$ and levels of B sum to $[B]$. Therefore, $\nu_3 = 1$. Finally, note that there are no degrees of freedom for the saturated model (i.e., $\nu_4 = 0$) since all elementary cell frequencies are completely determined by the observed sample outcome.

The foregoing intuitive approach for determining ν can be extended for use with two-dimensional tables having more than four cells. (It can even be used with tables of greater dimensionality.) But such an approach becomes increasingly cumbrous and prone to error as the size and dimensionality of tables become large. Obviously, a rule or algorithm to assist in this determination would be desirable, and such a rule will be presented in the next chapter. For now, since the values of residual χ^2 and L^2 are known, and since we also know ν for each test, the probability of observing each respective test statistic can be routinely assessed. These p-values will be presented in Table 4.2.

Component Chi-square Comparisons

In addition to the residual chi-square, a component L^2 value may be computed for all models. Because a component L^2 value is the difference between two residual L^2 statistics, these components can be used to compare models arranged in pairs. For example, to compair the relative goodness-of-fit of Model (4) in relation to Model (3), a component L^2 is calculated by subtracting the residual L^2 observed for Model (4) from the residual L^2 found for Model (3). For the working example, the component L^2 *associated with Model (4)* is:

$$L^2_{3-4} = L^2_3 - L^2_4 = 5.19 - 0.00 = 5.19$$

This component can be used to judge whether Model (3) should be chosen over Model (4). Granted, since Model (3) contains one less term than Model (4), Model (3) will probably not fit observed data as well as Model (4). But how much of a loss in ability to fit data would result from choosing Model (3) instead of Model (4)? The answer, in part, is provided by L^2_{3-4}, which indicates that the adoption of Model (3) will result in an increased accumulation of 5.19 units of residual chi-square; and the greater the increase in residual chi-square, the poorer the fit. So by selecting Model (3) in favor of Model (4), one indeed sacrifices some ability to fit observed data. But then, if the sacrifice is slight, the advantages of being able to advance a more parsimonious explanation—that offered by Model (3)—will outweigh the slight loss of ability to fit observed data accurately. We will discuss this point at greater length in the next section. For now, we compute the remaining pairwise component L^2's. Respectively, the component L^2's for Models (3) and (2) are:

$$L^2_{2-3} = L^2_2 - L^2_3 = 21.65 - 5.19 = 16.46$$
$$L^2_{1-2} = L^2_1 - L^2_2 = 25.68 - 21.65 = 4.03$$

Obviously, the greatest relative reduction in ability to fit observed elementary cells will occur should Model (2) be selected over Model (3).

Tests of Significance.   It is most fortuitous that component L^2's can be subjected to tests of statistical significance. The number of degrees of freedom for a given test are given by the difference in number of degrees of freedom for respective residual chi-square tests. For example, the component chi-square for the saturated model ($L^2_{3-4} = 5.19$) is tested on a single degree of freedom since $v_3 - v_4 = 1 - 0 = 1$. The results of tests of significance on component L^2's are also summarized in Table 4.2.

TABLE 4.2. Summary of Log-Linear Analysis of Data Presented in Table 1.1

Model	Fitted Marginals	Residual			Component		
		L^2	df	p	L^2	df	p
(1)	n	25.68	3	.001			
(2)	$[A]$	21.65	2	.001	4.03	1	.050
(3)	$[A], [B]$	5.19	1	.025	16.46	1	.001
(4)	$[AB]$	0.00	0	1.000	5.19	1	.025

The Additive Properties of L^2. The additivity of the likelihood-ratio chi-square enables us to make direct comparisons of models when they are arranged in pairs. When arranged hierarchically, component L^2's even permit us to assess the specific contribution made by each model parameter to the fit of observed data. To appreciate how this is typically done, consider the completely null model which has a residual L^2 equivalent to 25.68. Now, recall that this residual value reflects the magnitude of departure of observed f_{ij}'s about a fitted table where each cell of the fitted table contains an expected frequency of 25 (i.e., $F_{ij} = \tau = 25$). Because L^2's are additive, the residual L^2 and ν for Model (1) can be partitioned into components linked to specific hierarchical models as shown below:

Model	Likelihood-ratio	df
(2)	$L_{1-2} = 4.03$	$\nu_{1-2} = 1$
(3)	$L_{2-3} = 16.46$	$\nu_{2-3} = 1$
(4)	$L_{3-4} = 5.19$	$\nu_{1-2} = 1$
(1)	$L_1^2 = 25.68$	$\nu_1 = 3$

As we see, the component L^2's of Models (2), (3), and (4) sum to the residual chi-square for the completely null model, and the numbers of degrees of freedom manifest corresponding additivity.

Component L^2's enable us to identify terms in the model that, in a relative sense, appear not to be important. Relatively unimportant terms are those that contribute little to the reduction of residual chi-square. Or, from another perspective, components enable us to identify specific terms that appear to be important: terms that have the effect of markedly improving the fit between expected and observed cell frequencies, as evidenced by prominent component values. Since this concept is often difficult to grasp initially, several concrete examples are offered.

As a starting point, consider the fit (or lack of fit) provided by the single-parameter completely null model, $L_1^2(3) = 25.68, p < .0001$. By adding τ_i^a to this model, which takes into account observed differences in $[A]$, the residual departure is reduced by 4.03 units. Moreover, as shown in Table 4.2, this reduction of the residual (or improvement in fit) is significant at the .05 level. (If the data analyst is working under an a priori alpha level of .05, then factors responsible for the significant difference in the main-marginals of Variable A should be given serious consideration.)

Let us now add τ_j^b to the model to achieve Model (3), and a further reduction of residual departure. Specifically, through the additional fitting of $[B]$, we see a highly significant reduction of 16.46 residual units ($p < .001$). Finally, let us contrast the residual L^2's of Models (1) and (3). The resultant component here is $L_{1-3}^2 = 25.65 - 5.19 = 20.49$, which upon subjecting to test on two degrees of freedom (i.e., $\nu_{1-3} = \nu_1 - \nu_3 = 2$) is also found to be highly significant ($p < .001$). This particular component tells us that differences are present in $[A]$ or $[B]$, or both. Parenthetically, because of our previous work, we know that unequal distributions are present in both $[A]$ and $[B]$.

In sum, the likelihood-ratio component L^2's provide us with the means to determine which specific effects, or groupings of effects, are contributing significantly to the improvement of fit between expected and observed cell frequencies. This salutary feature turns out to be most helpful with an asymmetrical analysis, for it enables us to approach the analysis in a manner much like that seen in the ANOVA.

MODEL SELECTION

"Unfortunately," as Fienberg (1977) has stated, "there is no all-purpose, best method of model selection" (p. 47). Nevertheless, there is a guiding concept and two statistical procedures that, in the absence of strong a priori theory, generally lead to the selection of the most appropriate model. The guiding concept has been mentioned: it is that of *parsimony*. The "law of parsimony" is fundamental to scientific inquiry. The law, in brief, holds that when confronted with a choice between competing explanations (hypotheses or theories), the best explanation is that which adequately explains existing data in the simplest terms. It is a sound law. When applied to our work, this means that the most desired, most acceptable explanation is that provided by the contingency table model that contains the fewest terms, yet still fits observed data reasonably well. To assist in the selection of this model, both residual and component chi-square procedures are used.

Residual chi-squares are used to exclude from serious consideration unsaturated models that obviously do not fit observed data well. Referring to Table 4.2, both Models (1) and (2) can be eliminated immediately because of their highly significant L^2 values. The disposition of Model (3) depends upon the level of significance that the researcher has adopted, cognizant of sample size, as the criterion for exclusion. Adoption of the .05 level, for example, would lead to the exclusion of Model (3). But if the .01 level of significance was the criterion, Model (3) would be retained, although the fit provided by this model is suspect.

Following the use of residual chi-squares to eliminate poor-fitting models, component L^2's can be used to make determinations among remaining models. As we have learned, when hierarchical models are examined in pairs, the magnitude of the component L^2 belonging to the more complete model indicates the loss of fit that would be realized should the simpler model be selected. The examination usually begins with the saturated model and proceeds in a "step-down" fashion until a component L^2 is observed that is decidedly statistically significant. When encountered, we proceed no further because the selection of a model with fewer terms than the model manifesting the significant component L^2 would result in a significant loss of fit.

To apply this strategy to the models of Table 4.2, we start at the bottom of the table and assess the component for the saturated model, i.e., $L^2_{3-4} = 5.19$, $p < .025$. Whether we move upward to simpler models depends on the chosen level of significance and on considerations rooted in the substance of the research problem. If the decision is to be made solely by statistical criteria and it happens that the a priori alpha level was established at .05, then L^2_{3-4} would be deemed statistically significant and the saturated model would be selected. After all, proceeding to Model (3) would result in a significant ($p < .05$) loss in ability to fit observed data. On the other hand, if the a priori alpha level had been .01, the researcher would proceed to examine the component for Model (3). Since this component is highly significant, $L_{3-2}^2(1) = 16.46$, $p < .001$, further attempts to find a model would be terminated and Model (3) would be selected.

The reader may suspect (and correctly so) that the hypothetical data of Table 1.1 were deliberately constructed to yield an intricate result. Intricate results occur frequently in log-linear work, and we should, therefore, begin to appreciate the fact that model selection is often as much an art as it is a science. Further, the failure of the working example to provide us with a clean "textbook" result provides us with an excellent opportunity to mention briefly several additional considerations pertaining to the art of model selection.

If, as above, conventional statistical criteria do not point unequivocally to an acceptable model, it should go without saying that the researcher should select from among serious competitors the model that possesses the greatest subject-matter relevancy. Keep in mind always that statistical tests are merely tools—tools to assist, but not to replace, substantive judgement. Unfortunately, it is most difficult to illustrate how substantive concerns would lead to the selection of a model for our working example because our example has no history, data are not real, and there has been little or no intellectual investment in the problem. When dealing with an actual problem and real data, however, the researcher will find himself or herself in a better position to ensure that the statistical tail is not permitted to wag the substantive dog.

Conceding that subject matter relevancy is the ultimate criterion, there are, nevertheless, some additional considerations that often prove to be of value. Tests on the values of specific parameters in the model constitute one such consideration. Suppose that neither chi-square procedures (residual and component) nor substantive considerations pointed to an acceptable model in the working example. In other words, we are still faced with a choice between Model (3) and the saturated model where the importance of the four λ_{ij}^{ab} parameter estimates is central to an intelligent decision. One approach to the decision would be to perform an analysis of residuals (see Chapter 3) about the independence model. If residuals should be found to be statistically significant, the saturated model would clearly be the model of choice. Another approach, similar to residual analysis in simple situations, is to subject interaction-effect parameters to statistical test. It will be shown that these estimated effect parameters are distributed normally under the null hypothesis $H_o: \lambda_{ij}^{ab} = 0$, and that the standard error for a unit normal test can be obtained.

Consequently, these interaction effects can be tested for statistical significance with a standard z test. If these parameter estimates are found to be nonsignificant (at, say, the .05 level), justification is afforded to acceptance of Model (3). Conversely, should significance be achieved, the saturated model would appear to be the model of choice. We will have occasion to test parameters in this manner in the future.

MODEL INTERPRETATION

As we know, there is a different omnibus conclusion associated with each model for each of the two modes of inquiry (see Table 4.1). If Model (3) should be chosen as most acceptable, for a symmetrical analysis one would conclude that Variables A (Sex) and B (Attitude) are not associated. The conclusion associated with Model (3) in the asymmetrical case is that there are no differences between females and males with respect to their respective responses to the abortion issue. Realize that these conclusions constitute reasonable explanations, explanations most consistent with data, and not the logical alternatives to a rejected null hypothesis.

Should Model (4) be selected, then because λ_{ij}^{ab} is needed in the model to achieve reasonable fit, observed cell frequencies are obviously deviating from cell frequencies expected under the hypothesis of variable independence. Consequently, A and B are associated. If the inquiry is asymmetrical, prominent main effects exist between levels of the explanatory variable, i.e., females and males differ in their response to the abortion-issue variable.

POST HOC FOLLOW–UP PROCEDURES

Understandably, methods for following-up an omnibus result for complex two-dimensional tables and tables of higher dimensionality lack full development at this time. But following-up initial results emanating from fourfold tables is relatively straightforward. Of course, follow-up procedures for fourfold tables need only be considered if the saturated model is selected. If this model is adopted for a symmetrical analysis, the direction (positive or negative) of the association can be determined by inspection, and the intensity of the relationship can be assessed by *measures of association* such as a ratio of conditional odds or an adaptation of this ratio known as Yule's Q. For an asymmetrical analysis, procedures parallel to those used subsequent to an ANOVA can be highly informative. Specifically, the magnitude and direction of relevant lambda (or tau) parameters can be examined and tested for statistical significance. We will have a number of opportunities to pursue the results of an asymmetrical analysis in later chapters.

Measures of Association

Assuming an unbiased sample and compliance with the requisite conditions outlined in Chapter 3, acceptance of the saturated model simply means that the two qualitative variables are not independent. In a sense, acceptance of the saturated model is analogous to approaching a busy intersection and getting a green light to proceed with additional investigations. When the mode of inquiry is symmetric, either variable can be considered the response variable, thus additional investigations center not on *effects* (group differences) but rather on the *form* and the *strength* of the association. The form or nature of the relationship is easily determined by the "inter-ocular" test, a most under-rated and under-utilized statistical procedure. Simply put, for the working example, one visually contrasts the expected frequencies given by the saturated model (the f_{ij}'s) with the F_{ij}'s given by Model (3), the independence model. Before going any further, the reader is asked to perform this "test" on data in Table 3.2. Assuming compliance to this request, we conclude that females tended to assume a more negative posture toward the antiabortion amendment than did males; or, equally as valid, of those opposed to the amendment, females were represented to a greater degree than were males.

A most encouraging trend among behavioral researchers is the growing awareness that the "practical" significance of a relationship (or effect) is not provided by merely reporting a statistically significant test statistic with its associated p value. After all, the magnitude of a test statistic such as the L^2, and hence its p value, is to a considerable extent a function of sample size. In fact, to push this point to its extreme, it can be said that any null hypothesis can be statistically rejected if sample n's are sufficiently large. Consequently, subsequent to documenting a statistically significant relationship, it is often desirable to assess the *intensity* of the relationship independent of sample size. And as mentioned in Chapter 3, a number of measures of association for contingency tables have been proposed for this purpose. It is not our intent here to examine all of these measures, for this has been done well elsewhere (e.g., Reynolds, 1977). Two measures of association, however—the odds ratio and Yule's Q—have particular relevance in log-linear work. Although old friends of sociologists, these measures are relatively unfamiliar to social workers, psychologists, and the like; hence they merit, if only briefly, our attention.

The Odds Ratio. A significant χ^2 test statistic means that certain intratabular cell proportions differ. One way to view these differences is to first fix upon one of the variables. Then, for each level of that variable, compute the odds of responding to a specific level of the other variable (i.e., compute the conditional odds). Finally, compare the conditional odds over levels of the fixed variable. In fourfold tables, this procedure is relatively simple and yields a most meaningful measure of association called the *ratio of conditional odds* or, for short, the odds ratio.

Consider yet again the data in Table 1.1 and momentarily fix on Variable A; that is, consider it to be explanatory. Consider specifically the A_1 condition (females) where the *conditional odds* of opposing the antiabortion amendment are $f_{11}/f_{12} = 33/7 = 4.71$ or almost 5 to 1. For males, however, the odds of responding negatively are $f_{21}/f_{22} = 37/23 = 1.61$, or only about $1\frac{1}{2}$ to 1. Now, does it not follow that if there were no association between Sex and Attitude, the odds of opposing the amendment would be essentially the same for both females and males? By similar argument, to the extent to which the two variables are associated, the conditional odds will differ. Therefore, a comparison of respective conditional odds should yield a measure of the strength of the association.

The desired comparison, a ratio of conditional odds, is given by

$$\Omega_{1/2} = \frac{f_{11}/f_{12}}{f_{21}/f_{22}} \tag{4.22}$$

As seen, the numerical value of the odds ratio is symbolized by an uppercase Greek omega. Here, omega also carries the subscript 1/2 to reflect the *direction* of the conditional odds which, for the moment, are the odds of belonging to B_1 as opposed to B_2. If qualitative variables are independent, that is, if females and males have precisely the same odds of being opposed to the amendment, then the value of Ω will be 1. Conversely, deviation from unity in either direction is suggestive of an association between variables.

Incidentally, should we multiply both the numerator and denominator of the ratio in Eq. (4.22) by the quantity (f_{12}/f_{21}), upon simplification we would find that

$$\Omega_{1/2} = \frac{f_{11}/f_{12}}{f_{21}/f_{22}} = \frac{f_{11}/f_{21}}{f_{12}/f_{22}} \qquad (4.23)$$

which demonstrates that the odds ratio is invariant with respect to the variable chosen as explanatory.

Exercising Eq. (4.22) on our working-example data gives an odds ratio that is deviant from unity:

$$\Omega_{1/2} = \frac{33/7}{37/23} = \frac{4.714}{1.609} = 2.93$$

As an additional exercise, let us reverse the direction of the conditional odds. That is, let us consider the odds of being classified in B_2 (or A_2) as opposed to B_1 (or A_1). Computing the value of $\Omega_{2/1}$ results in

$$\Omega_{2/1} = \frac{37/23}{33/7} = \frac{37/33}{23/7}$$

$$= \frac{1.609}{4.714} = \frac{1.121}{3.286} = .34$$

which, we should note, is the reciprocal of $\Omega_{1/2}$. Thus, although the strength of the association between Variables A and B has not changed, by reversing the direction of the relationship the value of the odds ratio here turns out to be less than unity instead of greater than unity.

As a measure of association, Ω possesses both strengths and weakness. A principal strength is that the magnitude of Ω is independent of sample size, a fact that enables us to compare the strength of two or more associations computed on samples of different sizes. Also, just as desirable, the magnitude of Ω is not artificially reduced in the presence of unequal marginal distributions. Recall that in Chapter 3 it was noted that the values of many measures of association (e.g., the phi coefficient and the contingency coefficient) only achieve their maximum when the association is perfect and the marginal distributions within [A] and [B] are even. To the extent to which the latter condition does not hold, the maximum value of these measures is suppressed—just as the maximum value of a Pearson-product-moment correlation is suppressed when variable distributions depart from normality. Fortunately, this limiting feature does not occur with the odds ratio.

The odds ratio, however, is not without its problems. Surely the reader experienced some discomfort a moment ago when the intensity of the association of the working example was first computed as $\Omega_{1/2} = 2.93$ and, subsequent to reversing the directionality, then as $\Omega_{2/1} = .34$. The source of the discomfort is that the underlying metric of the odds ratio is partially unbounded and, to compound problems of interpretation, the metric is not symmetric on a unit scale about unity. Granted, the substantive center of the metric is anchored about 1, indicating no relationship; but as the association gains in intensity, the value of Ω approaches either 0 or extends to infinity (∞), depending upon the arbitrary choice of direction. In other words, the underlying metric occupies the entire range of the domain of nonnegative real numbers but it is not symmetric about 1, its center. (Between 0 to unity, the metric spans only one interval unit while the metric between unity and ∞ occupies an infinite number of units.) Clearly, interpretation of this measure would be greatly enhanced if the lower and upper boundaries could be harnessed and if the metric could be made symmetric about a point representing an absence of association. This has been done, at least for 2 X 2 tables, by Yule (1900).

Yule's Q Statistic. The Q statistic is a function of a rather well-known measure called the *crossproduct ratio* (CPR) which in turn is a function of the odds ratio. To appreciate these relations, we start with the odds ratio formula of Eq. (4.22) and then simplify this ratio by "inverting the denominator" and multiplying the inverted denominator by the original numerator. The simplified results shown on the extreme right of Eqs. (4.24) and (4.25) below are the crossproduct ratios for fourfold tables:

$$\Omega_{1/2} = \frac{f_{11}/f_{12}}{f_{21}/f_{22}} = \frac{f_{11}f_{22}}{f_{21}f_{12}} \tag{4.24}$$

$$\Omega_{2/1} = \frac{f_{12}/f_{11}}{f_{22}/f_{21}} = \frac{f_{12}f_{21}}{f_{22}f_{11}} \tag{4.25}$$

Note that the CPR constitutes a more economical way of computing an odds ratio. Also, Yule found that by (a) subtracting 1 from the CPR, (b) adding 1 to the CPR, and (c) forming a ratio of these reduced and augmented quantities, a most meaningful measure of association is obtained. This measure, Yule's Q, is developed more fully below:

$$Q_{1/2} = \frac{\Omega - 1}{\Omega + 1} = \frac{\text{CPR} - 1}{\text{CPR} + 1}$$

$$= \frac{(f_{11}f_{22}/f_{12}f_{21}) - (f_{12}f_{21}/f_{12}f_{21})}{(f_{11}f_{22}/f_{12}f_{21}) + (f_{12}f_{21}/f_{12}f_{21})}$$

$$= \frac{(f_{11}f_{22}) - (f_{12}f_{21})}{(f_{11}f_{22}) + (f_{12}f_{21})} \qquad (4.26)$$

For the working example and from the perspective of being in the first level of B (or A), $Q_{1/2}$ is

$$Q_{1/2} = \frac{2.93 - 1}{2.93 + 1} = \frac{(33 \cdot 23) - (7 \cdot 37)}{(33 \cdot 23) + (7 \cdot 37)} = .49$$

From the opposite perspective, that of being in the second level of A or B, $Q_{2/1}$ is

$$Q_{2/1} = \frac{.34 - 1}{.34 + 1} = \frac{(7 \cdot 37) - (33 \cdot 23)}{(7 \cdot 37) + (33 \cdot 23)} = -.49$$

Yule's Q ranges in value from -1.00 to $+1.00$ and is symmetric, as demonstrated above, about zero. As Q approaches zero the association becomes weaker; as Q approaches unity, in either direction, the association increases in intensity. Although not without faults (see Reynolds, 1977, p. 26), Yule's Q has much to recommend it as a measure of association. And although its use is restricted to four-fold tables, the odds ratio which serves as the basis for this statistic can be generalized to tables of greater size and dimensionality as will be seen later.

Examination of Lambda Effects

As noted earlier, specific-effect parameters can be tested for significance in a manner similar to the testing of residuals. In fact, for two-dimensional tables, there is essentially little difference in practice between the testing of residuals about Model (3) and the testing of interaction effects. The latter approach can be extremely useful in more complex situations, however. Moreover, the testing of parameter effects can prove to be a most valuable follow-up strategy for asymmetrical analyses. Again, realize that for two-dimensional tables, follow-up is required only if the saturated model is selected. Further, for fourfold tables, extensive post hoc examination is not required, because there are only two levels of the explanatory variable (say Variable A) and acceptance of the saturated model suggests that the pattern of proportional response to Variable B is not similar in the two levels of A. Consequently, only the algebraic signs of λ_{ij}^{ab} effects need be examined to determine the nature of differences between levels of A.

To illustrate, recall that calculations of interaction effects, performed earlier in this chapter, turned out as follows:

| | \multicolumn{2}{c}{τ} | \multicolumn{2}{c}{λ} |
	B_1	B_2	B_1	B_2
A_1	1.308	.764	.269	−.269
A_2	.764	1.308	−.269	.269

Because the lambdas are slightly easier to interpret, they will be chosen for the ensuing discussions of post hoc procedures. From the algebraic signs of the lambdas, it can be seen that proportionately more females give a B_1 response—they are more disposed to be against the antiabortion amendment. Even though this uncomplicated result does not require further statistical scrutiny, as a pedagogic exercise we pursue our follow-up examination by subjecting the λ_{ij}^{ab} to statistical test.

In general, the null to be subjected to test is $H_o: \lambda_{ij}^{ab} = 0$, for all values of i and j. But since there is only one *basic parameter* here (i.e., when a solution is achieved for one, the remaining three effects are determined), in practice only one statistical test need be performed. With relatively large samples (e.g., $n > 25$), under the null hypothesis, the distribution of lambdas approaches the normal; hence, subsequent to calculating the standard error of the λ's, a conventional normal deviate test (i.e., z test) can be performed (Goodman, 1970). The standard error for the saturated model of a fourfold table may be computed as follows:

$$\text{S.E.}(\lambda) = \frac{\left[\sum_{i}^{a} \sum_{i}^{b} (1/f_{ij}) \right]^{1/2}}{ab} \qquad (4.27)$$

TABLE 4.3. Observed $[AB]$ Frequencies, Lambda Parameters, and Normal Deviate Tests for the Working Example

Attitude: Variable B	Sex: Variable A					
	Females			Males		
	f_{1j}	λ^{ab}_{1j}	z	f_{2j}	λ^{ab}_{2j}	z
Opposed	33	.269	2.19*	7	−.269	−2.19
Support	37	−.269	−2.19	23	.269	2.19
	70	0.000		30	0.000	

*$p < .05$.

Substituting values peculiar to our examples give

$$\text{S.E.}(\lambda) = (1/33 + 1/37 + 1/7 + 1/23)^{1/2}/(2 \cdot 2)$$
$$= (.2437)^{1/2}/4 = .123$$

Thus, the z test on an unspecified lambda is

$$z(\lambda^{ab}_{ij}) = \lambda^{ab}_{ij}/\text{S.E.}(\lambda) \qquad (4.28)$$

which for λ^{ab}_{11} gives us

$$z(\lambda^{ab}_{11}) = .269/.123 = 2.19$$

Since this computed test statistic exceeds the two-tailed critical z value of 1.96, the null that $\lambda^{ab}_{11} = 0$ may be rejected at the .05 level of significance. Further, since there is only one basic parameter, additional testing would be redundant. A summary of our post hoc examination as it might appear in a journal article is presented in Table 4.3.

CONCLUDING REMARKS

At the outset, we said that the traditional chi-square procedures of Chapter 3 are generally sufficient for the analysis of two-dimensional tables. If researchers typically possessed information on only two qualitative variables, it would be difficult to justify the new approach ponderously outlined in this chapter for, in retrospect, all we have done is make more explicit that which is implicit in the traditional approach. But since researchers often deal with more than two variables, and since the principles that were at times extensively deliberated in this chapter can be extended to situations in which there are more than two variables, our work here is justified. There is, of course, still much to learn about the application of log-linear analysis to two-dimensional tables. We could, for example, extend our methods to analyze data in two-way tables defined by two polytomous variables. However, it is advantageous to proceed immediately to the analysis of three-dimensional tables so that the full benefits of this relatively new generalized approach to the analysis of qualitative data can be more fully seen.

5

The Analysis
of Three-dimensional
Tables

Until recently, researchers who were able to cross-classify subjects on more than two variables and who desired to perform an analysis that considered all variables simultaneously were presented with serious problems, problems that were only partially ameliorated by the intricate partitioning methods proposed by Lancaster (1951). However, with the advent of log-linear theory and supporting computer programs such as BMDP/4F (Dixon, 1981) and ECTA (Fay & Goodman, 1975), contemporary researchers can readily perform comprehensive analyses of higher-order tables. This chapter describes the application of log-linear methods to tables of three dimensions. Adhering to our earlier practice, log-linear methods will be described in a manner synonymic with multifactor ANOVA. Also, as before, basic principles will be illustrated within the context of a contrived example (The Adverse Impact Study), but unlike before, refinements will be discussed within the context of an actual experiment (The Reflective Teaching Study).

We turn first to a consideration of a new working example, a study of "adverse impact" with respect to employment in an industrial setting. Our new example will serve to illustrate typical problems that are amenable to log-linear analysis, and it will provide us with data, albeit hypothetical data, to refine further basic principles. By and large, we will consider again the major operations surveyed in the last chapter, but in somewhat greater depth and with somewhat greater emphasis on the distinction between the general application of log-linear models for symmetrical inquiry and the application of a specific subset of models, called *logit models*, that are used when the mode of inquiry is asymmetrical. Finally, to reinforce principles, an actual example from the field of education will be presented.

THREE–DIMENSIONAL TABLES

A Study of Adverse Impact

Some time ago, this author was approached by a large distributing firm for assistance with their efforts to assess the outcome of a testing program designed to select qualified hourly employees for positions as forklift operators. The positions were newly created, the openings were numerous, and the financial remuneration was most attractive. Assistance was sought because management was concerned that its multiphase testing program might screen out a disproportionate number of female and minority applicants, thus rendering the company vulnerable to the charge that their testing and selection procedures were having an "adverse impact" on these particular groups. Hence, in addition to satisfying the criteria proposed by several federal agencies, management desired a more rigorous assessment of the testing program. Since the variables of interest were three in number and qualitative in nature, a log-linear analysis was deemed appropriate. Specifically, an asymmetrical logit-model analysis was proposed and subsequently implemented.

Briefly, employees who volunteered to participate in at least one phase of the testing program were cross-classified on the basis of the following three dichotomies:

A: Sex	B: Ethnic Group
A_1 = females	B_1 = majority
A_2 = males	B_2 = minority

C: Outcome at Termination of Testing
C_1 = failed to qualify for a position
C_2 = qualified for a position

TABLE 5.1. Observed Frequencies by Sex, Ethnic Group, and Test Performance for the Adverse Impact Study

		Test Performance	
Sex	Ethnic Group	C_1: Failed	C_2: Passed
A_1: Females	B_1: Majority	7	4
A_1: Females	B_2: Minority	3	6
A_2: Males	B_1: Majority	18	24
A_2: Males	B_2: Minority	12	26

Both to respect the confidentiality of actual outcome and to provide us with a convenient set of data with which to work, let us assume that 100 employees participated in the testing program ($n = 100$) and that cross-tabulations turned out as shown in Table 5.1.

Notation for Three-dimensional Tables

We can simply extend the system established in Chapters 3 and 4 to accommodate the introduction of the third and new variable, Variable C. Thus, under either a fixed or random sampling plan, n sampling units are cross-classified on the basis of three variables, Variables A ($i = 1, 2, \ldots, a$), B ($j = 1, 2, \ldots, b$), and C ($k = 1, 2, \ldots, c$). Observed elementary cell frequencies in the $[ABC]$ configuration are denoted by f_{ijk}'s; therefore, expected cell frequencies fitted by a given model will be symbolized by F_{ijk}'s. Observed main-marginals and two-variable (first-order) marginals are designated as follows:

Main-Marginals	Two-Variable Marginals
$[A]$: f_i^a	$[AB]$: f_{ij}^{ab}
$[B]$: f_j^b	$[AC]$: f_{ik}^{ac}
$[C]$: f_k^c	$[BC]$: f_{jk}^{bc}

Applying this system to the study of adverse impact, for the main-marginals we have

$[A]$: $f_1^a = 20$ and $f_2^a = 80$

$[B]$: $f_1^b = 53$ and $f_2^b = 47$

$[C]$: $f_1^c = 40$ and $f_2^c = 60$

For the two-variable marginals (or first-order marginals), we observe

[AB]			[AC]			[BC]		
A	B	f^{ab}	A	C	f^{ac}	B	C	f^{bc}
1	1	11	1	1	10	1	1	25
1	2	9	1	2	10	1	2	28
2	1	42	2	1	30	2	1	15
2	2	38	2	2	50	2	2	32

Finally, realize that each of the 100 employees is classified into one, and only one, of the eight elementary cells of the 2 × 2 × 2 table.

HIERARCHICAL MODELS FOR THREE-DIMENSIONAL TABLES

The structural synonymity between conventional ANOVA models and log-linear models was discussed in the preceding chapter. That discussion, however, was confined to tables defined by two qualitative variables. Relative to three-variable situations, the ANOVA analogue is a model that explains elementary cell *means* (i.e., μ_{ijk}'s) by a linear combination containing (a) a grand mean parameter μ, (b) main effects of each of the three principal variables, (c) three first-order interaction effects, and (d) a second-order (or three-variable) interaction. That is, the ANOVA analogue for contingency-table models is

$$\mu_{ijk} = \mu + \alpha_i + \beta_j + \gamma_k + (\alpha\beta)_{ij}$$
$$+ (\alpha\gamma)_{ik} + (\beta\gamma)_{jk} + (\alpha\beta\gamma)_{ijk} \tag{5.1}$$

The grand mean μ, the main effects for Variables A and B, and the first-order interaction between A and B have been defined previously. Moreover, from discussions associated with Eq. (4.1) the reader can readily deduce the definitions for the main effects of Variable C, represented by a lowercase Greek gamma (i.e., γ_k), and the two first-order interactions which involved Variable C. The last term in Eq. (5.1), however, merits special attention because the meaning of second-order interaction is not universally appreciated within the context either of ANOVA or of log-linear analysis.

In an ANOVA, the presence of second-order interaction indicates that elementary cell means manifest variability subsequent to correcting these means for main effects of Variables A, B, and C and subsequent to correcting the means for the three first-order interactions. Or put another way, the variability among cell means cannot be totally explained by the additive influences exerted by single variables (i.e., the main effects) or variables in dual combination (first-order interaction effects); rather, there exist other sources of cell-mean variability that are detected only when the three principal variables act in concert. The presence of second-order interaction can also be interpreted to mean that there is a change in the *nature* (or pattern) of *simple* first-order interactions if they are viewed over levels of the third (or excluded) variable. For example, should we arbitrarily choose to examine the interaction between Variables A and B peculiar to each level of Variable C, the pattern of these simple AB interactions will manifest some degree of change over levels of Variable C. Consequently, if the first-order interaction between A and B is substantively meaningful, simple first-order interactions must be interpreted independently at respective levels of C. As we shall see, an analogous situation will present itself in log-linear work whenever the eight-parameter saturated model is selected for interpretation.

General Multiplicative Models

As before, multiplicative models will be discussed prior to their log-linear counterparts. The eight models, displayed hierarchically, are shown in Table 5.2.

Several observations can be made from an examination of Table 5.2. First, the numbering system used to identify models is consistent with the number of general parameters within each model. Second, proceeding from Model (1) to the saturated model, more observed marginal information is used to fit F_{ijk}'s and hence to estimate parameters. Model (1), for example, uses only n to generate common F_{ijk}'s, whereas the saturated model uses all resultant elementary cell frequencies. (The latter model, therefore, will generate F_{ijk}'s identical to observed f_{ijk}'s.) Finally, the eight factors in the saturated model are analogous to the eight terms of the ANOVA model presented in Eq. (5.1).

Before leaving Table 5.2, the reader should appreciate the linkage between the parameters within a model and the marginal information fitted by that model. The relevance of the "Fitted Marginals" column in Table 5.2 to both theory and computer application cannot be overstated. As a case in point, to access information about a specific model on most current computer programs, the model in question is specified by the marginal information used to fit F_{ijk}'s. For example, to examine the fit produced by Model (5) on most computer programs —e.g., the 4F program in the BMDP computer package or on ECTA (Everyman's Contingency Table Analysis)—desired results are obtained by requesting output for that model which fits both the $[AB]$ configuration and the main-marginals $[C]$. Note that for hierarchical models, the fitting of $[AB]$, the observed two-dimensional table defined by the crossing of Variables A and B, also provides for the fitting of its lower-order relatives $[A]$ and $[B]$. Consequently, requesting the fit of only $[AB]$ defines the model

$$F_{ijk} = \tau \tau_i^a \tau_j^b \tau_{ij}^{ab}$$

TABLE 5.2. Hierarchical Multiplicative Models for Three-dimensional Tables

Model Number	Multiplicative Models	Fitted Marginals
(1)	$F_{ijk} = \tau$	n
(2)	$F_{ijk} = \tau \tau_i^a$	$[A]$
(3)	$F_{ijk} = \tau \tau_i^a \tau_j^b$	$[A], [B]$
(4)	$F_{ijk} = \tau \tau_i^a \tau_j^b \tau_k^c$	$[A], [B], [C]$
(5)	$F_{ijk} = \tau \tau_i^a \tau_j^b \tau_k^c \tau_{ij}^{ab}$	$[AB], [C]$
(6)	$F_{ijk} = \tau \tau_i^a \tau_j^b \tau_k^c \tau_{ij}^{ab} \tau_{ik}^{ac}$	$[AB], [AC]$
(7)	$F_{ijk} = \tau \tau_i^a \tau_j^b \tau_k^c \tau_{ij}^{ab} \tau_{ik}^{ac} \tau_{jk}^{bc}$	$[AB], [AC], [BC]$
(8)	$F_{ijk} = \tau \tau_i^a \tau_j^b \tau_k^c \tau_{ij}^{ab} \tau_{ik}^{ac} \tau_{jk}^{bc} \tau_{ijk}^{abc}$	$[ABC]$

Note. To fit a particular model on many canned computer programs, the capital letters representing the marginals that are fitted by the model are specified, separated by commas, and terminated with a period. For example, to fit Model (6) using the BMDP/4F, one requests a fit for: *AB, AC*. In ensuing tables, therefore, the brackets will be omitted when specifying marginals that are fitted by particular models.

TABLE 5.3. Hierarchical Log-Linear Models for Three-dimensional Tables

Model Number	Log-Linear Models
(1)	$\ln F_{ijk} = \lambda$
(2)	$\ln F_{ijk} = \lambda + \lambda_i^a$
(3)	$\ln F_{ijk} = \lambda + \lambda_i^a + \lambda_j^b$
(4)	$\ln F_{ijk} = \lambda + \lambda_i^a + \lambda_j^b + \lambda_k^c$
(5)	$\ln F_{ijk} = \lambda + \lambda_i^a + \lambda_j^b + \lambda_k^c + \lambda_{ij}^{ab}$
(6)	$\ln F_{ijk} = \lambda + \lambda_i^a + \lambda_j^b + \lambda_k^c + \lambda_{ij}^{ab} + \lambda_{ik}^{ac}$
(7)	$\ln F_{ijk} = \lambda + \lambda_i^a + \lambda_j^b + \lambda_k^c + \lambda_{ij}^{ab} + \lambda_{ik}^{ac} + \lambda_{jk}^{bc}$
(8)	$\ln F_{ijk} = \lambda + \lambda_i^a + \lambda_j^b + \lambda_k^c + \lambda_{ij}^{ab} + \lambda_{ik}^{ac} + \lambda_{jk}^{bc} + \lambda_{ijk}^{abc}$

Note. Consult Table 5.2 for observed marginal configurations that are used for parameter estimation and model specification.

But to work with Model (5) in Table 5.2, a model that contains τ_k^c in addition to the four parameters shown above, [C] must also be fitted. To call for Model (6), [AB] and [AC] are fitted; and by that fact, the lower-order relatives [A], [B], and [C] are fitted also.

Log-Linear Models

From Chapter 4, we know that the multiplicative models of the preceding section are linear in their logarithms. Log-linear models parallel to those in Table 5.2 are displayed in Table 5.3. Again, notice that log-linear models produce expected cell frequencies that correspond to the natural logs of F_{ijk}'s produced by their multiplicative counterparts. And whereas linear models are more amenable to existing computational routines and are generally favored by behavioral researchers, realize that there is no "significant" difference between corresponding multiplicative and log-linear models.

INTERPRETING GENERAL MODELS IN SYMMETRICAL INQUIRY

In this section, the *general* log-linear models in Table 5.2 will be examined for the purpose of determining the omnibus conclusion that can be advanced should the model in question be deemed acceptable. The mode of inquiry will be assumed to be symmetrical;

conclusions associated with logit-models for asymmetrical inquiry will be discussed in the next section. The concrete example to be used is the study of adverse impact even though it is recognized that this study more readily lends itself to an asymmetrical logit-model analysis.

The didactic strategy adopted for the ensuing discussion is as follows:

1. Models will be successively identified and basic probability theory, where appropriate, will be used to generate expected elementary cell frequencies.

2. Expected cell frequencies for each model will be presented in tabular form.

3. We will momentarily assume that the expected cell frequencies fit observed data extremely well and that the model in question has been selected for interpretation.

4. A general interpretation or conclusion, in terms of symmetrical concepts, will be inferred.

5. For selected models, the numerical values of parameter estimates will be calculated.

Before we begin, however, it should be realized that the approach adopted for this section is not fully consistent with methods generally used by computer programs designed for use in log-linear work. Unfortunately, for certain models it is not possible to resort to elementary probability theory to generate expected cell frequencies. We will encounter this problem with Model (7). The problem is not an intractable one, however, for iterative routines have been developed that use ML procedures to mechanically generate expected cell frequencies for any model that fits a complete table. Such a procedure has been proposed by Deming and Stephan (1940), described by Fienberg (1970) and Goodman (1970), translated into Fortran by Haberman (1972), and illustrated for Model (7) of the adverse impact study in Appendix A. Finally, following the generation of expected cell frequencies, computer programs determine the numerical values of the effect parameters in the model (i.e., the taus or lambdas), and unless otherwise requested these values will be consistent with the definitions of effect parameters contained in this chapter.

Model (1): Mutual Equiprobability

The multiplicative version of this model contains only τ and uses only n to determine its value. Thus, if we "fit" n to a three-dimensional table, since there are no other constraints (i.e., parameters) in the model the n counts will be distributed uniformly throughout the cells of the table. Therefore, the F_{ijk}'s may be computed directly by

$$F_{ijk} = n/abc = 100/8 = 12.5 \qquad (5.2)$$

which, for the study of adverse impact, yields the following table of fitted cell frequencies:

A	B	C_1	C_2
1	1	12.5	12.5
1	2	12.5	12.5
2	1	12.5	12.5
2	2	12.5	12.5

Assuming that the fitted frequencies above describe observed data extremely well, what omnibus conclusion could be advanced relative to the symmetric relations between and among the three variables? It is evident that not only are the three variables mutually independent, i.e., $A \otimes B \otimes C$, but an equal number of cases falls into each cell. Hence, the adoption of Model (1) would connote *mutual equiprobability*.

Turning to the numerical value of the sole parameter, from Chapter 4 we learned that τ is analogous to μ, the grand mean in ANOVA models. More precisely, τ is equivalent to the geometric mean of F_{ijk}'s given by the model in question. Hence,

$$\tau = \bar{G} = \left(\prod F_{ijk} \right)^{1/abc} = [(12.5)^8]^{1/8} = 12.5$$

or, should the log-linear be desired,

$$\lambda = \ln \tau = \ln 12.5 = 2.5257$$

Model (2): Conditional Equiprobability

This two-parameter model fits $[A]$ to three-dimensional tables. In our example, $f_1^a = 20$ and $f_2^a = 80$. Hence, subsequent to the restriction that the main marginals for A must sum to the respective values above, counts within levels of A are independently or uniformly distributed. Expected cell frequencies, therefore, may be directly computed by

$$F_{ijk} = f_i^a/bc \qquad (5.3)$$

which, if carried through, yields the following fitted frequencies for the working example:

A	B	C_1	C_2
1	1	5	5
1	2	5	5
2	1	20	20
2	2	20	20

From the table it is clear that $A \otimes B \otimes C$. But, in addition, responses are equiprobable if observed differences in f_1^a and f_2^a are taken into account. Therefore, if Model (2) fits observed data well, *conditional equiprobability* best describes the omnibus conclusion. That is, responses are equally probable on the condition that adjustments are made for the observed inequalities in the main marginals of Variable A.

Let us calculate the values of the two basic parameters in this model. As always, the first parameter is the geometric mean of fitted F_{ijk}'s. Consequently

$$\tau = \bar{G} = (5 \cdot 5 \cdot 5 \cdot 5 \cdot 20 \cdot 20 \cdot 20 \cdot 20)^{1/8} = 10.00$$

And, in the log-linear, the first parameter is

$$\lambda = \ln \tau = \ln 10.00 = 2.3026$$

Now, recall that the generic expression for the second parameter is

$$\tau_i^a = \bar{G}_i^a/\bar{G}$$

and so the specific effects are

$$\tau_1^a = \bar{G}_1^a/\bar{G} = (5 \cdot 5 \cdot 5 \cdot 5)^{1/4}/10.00 = .50$$
$$\tau_2^a = \bar{G}_2^a/\bar{G} = (20 \cdot 20 \cdot 20 \cdot 20)^{1/4}/10.00 = 2.00$$

Note that $\Pi\tau_i^a = 1.00$; hence, for dichotomies, $\tau_1^a = 1/\tau_2^a$ or $\tau_2^a = 1/\tau_1^a$. For readers familiar with ANOVA, more insightful definitions of these effects are provided by the linear expressions

$$\lambda_1^a = \ln \bar{G}_1^a - \ln \bar{G} = 1.6094 - 2.3026 = -.6932$$

$$\lambda_2^a = \ln \bar{G}_2^a - \ln \bar{G} = 2.9957 - 2.3026 = \quad .6932$$

The "ANOVA-like" effects above sum to zero as do fixed effects in the analysis of variance, which should not come as a complete surprise because the log of the product of multiplicative effects (i.e., unity) is zero.

Operational versions of Model (2) for the adverse impact study are

$$F_{1jk} = (10)(.50) \qquad F_{2jk} = (10)(2.00)$$

Exercising these models for all values of j and k will yield the expected cell frequencies shown earlier.

Model (3): Conditional Equiprobability

The conditional equiprobability model fits both $[A]$ and $[B]$. If we were dealing with a two-dimensional table, F_{ij}'s would be the products of $np_i^a p_j^b$. But we are working with a three-dimensional table and, since the third variable (Variable C) is not fitted, frequencies will be evenly distributed over this variable. It follows, therefore, that

$$F_{ijk} = np_i^a p_j^b / c \qquad\qquad (5.4)$$

The expected cell frequencies are

A	B	C_1	C_2
1	1	5.3	5.3
1	2	4.7	4.7
2	1	21.2	21.2
2	2	18.8	18.8

Not only do we see that $A \otimes B \otimes C$, but additionally we observe another instance of *conditional equiprobability*—equal probabilities over levels of Variable C following adjustments for the main-marginals of Variables A and B. Relative to the three basic parameters in this model, the reader is encouraged to calculate these values. The values should turn out to be:

$$\tau = 9.982$$

$$\tau_1^a = .500 \qquad \text{hence } \tau_2^a = 2.000$$

$$\tau_1^b = 1.062 \qquad \text{hence } \tau_2^b = .942$$

Notice that the numerical value of τ has changed from that computed for Model (2) but the main-marginal effects for Variable A are as they were in Model (2).

Model (4): Mutual Independence

Since the mutual independence model contains all three main-marginal parameters, it is often referred to as the *full* main-marginal model. Either of the expressions to the right in Eq. (5.5) below can be used to fit $[A]$, $[B]$, and $[C]$:

$$F_{ijk} = np_i^a p_j^b p_k^c = f_i^a f_j^b f_k^c / n^2 \tag{5.5}$$

The resultant frequencies are:

A	B	C_1	C_2
1	1	4.24	6.36
1	2	3.76	5.64
2	1	16.96	25.44
2	2	15.05	22.56

The fitted frequencies reflect a condition of *mutual independence*, i.e., $A \otimes B \otimes C$. Should Model (4) fit observed data extremely well, then there is no evidence of a relationship between any pair of variables. Finally, it turns out that $\tau = 9.780$, the τ_i^a's and τ_j^b's have the same values as seen for Model (3), and $\tau_1^c = .816$; hence, $\tau_2^c = 1.225$.

Model (5): Marginal Association between A and B

The marginal association between A and B model contains a first-order interaction parameter, namely τ_{ij}^{ab}. To fit this model, frequencies in $[AB]$ are used; consequently the main-marginals $[A]$ and $[B]$ are also fitted. Main-marginal $[C]$, however, must be fitted directly. This is accomplished by

$$F_{ijk} = f_{ij}^{ab} p_k^c \qquad (5.6)$$

which gives

A	B	C_1	C_2
1	1	4.4	6.6
1	2	3.6	5.4
2	1	16.8	25.2
2	2	15.2	22.8

If the incorporation of τ_{ij}^{ab} into the model is needed to achieve respectable fit, then not only are the three qualitative variables *not* mutually independent, but it can be said that there is a *marginal association* between Variables A and B. To be more precise, if we ignore Variable C by simply collapsing over its levels to form the $[AB]$ configuration, Variables A and B will manifest an association in this two-dimensional table. The intensity of the marginal relationship between A and B, of course, will depend on the extent to which τ_{ij}^{ab} is needed to achieve acceptable fit. Also, it is just as important to consider parameters that are not needed in the model (τ_{ik}^{ac} and τ_{jk}^{bc} in this case) since their omission suggests that *marginal independence* characterizes $[AC]$ and $[BC]$.

Model (5) offers us a good opportunity to reinforce our understanding of model parameter estimates. Consider first the geometric mean of F_{ijk}'s, which turned out to be 9.767 here (hence, $\lambda = 2.279$). Now it should have been noticed that since models generate different F_{ijk}'s, the first parameter is numerically different for each model discussed to this point. But what changes, if any, have occurred in main-marginal parameter estimates? For the present model, it can be shown that $\tau_1^a = .499$, $\tau_1^b = 1.078$, and $\tau_1^c = .816$. A comparison with earlier models reveals that the estimates associated with Variables A

and B have changed, but the τ_k^c's have not changed from Model (4). But this was to be expected since by fitting $[AB]$ for Model (5) we have effected a change in the value of $\bar{G}_i^a$'s and $\bar{G}_j^b$'s relative to Model (4); however, main-marginal effects for Variable C were not constrained by fitting $[AB]$, and hence these values are identical in both Models (4) and (5).

Let us take a hard look at the first-order effects for Model (5). The point was made in Chapter 4 that these effects are analogous to first-order interaction effects in the ANOVA. In that chapter, the analogy between interaction in the ANOVA (see Eq. 4.2) and contingency-table interaction (see Eqs. 4.16 and 4.21) was developed when τ_{ij}^{ab}, defined as

$$\tau_{ij}^{ab} = f_{ij}^{ab}/(\tau \tau_i^a \tau_j^b)$$

was converted to λ_{ij}^{ab} and shown to be

$$\lambda_{ij}^{ab} = \ln f_{ij}^{ab} - \lambda - \lambda_i^a - \lambda_j^b$$
$$= \ln f_{ij}^{ab} - \ln \bar{G}_i^a - \ln \bar{G}_j^b + \ln \bar{G}$$

Although the formulas above were developed within the context of two-dimensional tables, they can be extended for three-dimensional tables. Consider first the fact that the number of F_{ijk}'s in a combination (cell) formed by the crossing of Variables A and B is, in general, c. For the working example, $c = 2$. Thus it follows that we can extend the definition of first-order interaction between A and B to mean variability among the logs of the ab geometric means in $[AB]$ after partialling out the effects of $\ln \bar{G}_i^a - \ln \bar{G}$ and $\ln \bar{G}_j^b - \ln \bar{G}$. Stated mathematically, for three-dimensional tables,

$$\tau_{ij}^{ab} = \left(\prod_k^c F_{ijk} \right)^{1/c} /(\tau \tau_i^a \tau_j^b)$$
$$= \bar{G}_{ij}^{ab}/(\tau \tau_i^a \tau_j^b) \tag{5.7}$$

or,

$$\lambda_{ij}^{ab} = \ln \bar{G}_{ij}^{ab} - \ln \bar{G}_i^a - \ln \bar{G}_j^b + \ln \bar{G}$$

Using the multiplicative formula to compute the interaction effect for the first level of A and B in the working example, we find

$$\tau_{11}^{ab} = [(4.4)(6.6)]^{1/2}/[(9.767)(.499)(1.078)]$$

$$= 5.389/5.254$$

$$= 1.026$$

The corresponding linear estimate is

$$\lambda_{11}^{ab} = 1.684 - 1.584 - 2.354 + 2.269$$

$$= .025$$

Since, from our study of the last chapter, we know that marginal products of interaction effects in the [AB] configuration are unity (or marginal sums are all zeros in the linear), the remaining three interaction effects for the 2 × 2 table can be calculated most easily as residual quantities.

Model (6): Marginal Association between A and C

The F_{ijk}'s given by Model (6), a model that fits both [AB] and [AC] and thereby automatically fits the three main-marginals, may be calculated by hand by

$$F_{ijk} = np_{ij}^{ab}(p_{ik}^{ac}/p_i^a) \tag{5.8}$$

with the result

A	B	C_1	C_2
1	1	5.50	5.50
1	2	4.50	4.50
2	1	15.75	26.25
2	2	14.25	23.75

Assuming that τ_{ik}^{ac} is needed for respectable fit, then a *marginal* association is present between Variables A and C. (There may or may not be a marginal association between A and B.) In addition, it can also be said that Variables B and C are *conditionally independent* since τ_{jk}^{bc} is not part of the model. In other words, after adjusting for differences between levels of A, Variables B and C are not associated. Symbolically, this is expressed as $B \otimes C \mid A$.

Model (7): Full First-Order

Model (7) contains all three first-order parameters, and therefore it is sometimes called the *full* first-order model. It differs markedly from previous models in that direct calculation of F_{ijk}'s is not possible. Instead, as mentioned, expected cell frequencies are obtained by an iterative fitting procedure that is demonstrated in Appendix A. Iterative fitting yields the following:

A	B	C_1	C_2
1	1	6.288	4.712
1	2	3.712	5.288
2	1	18.712	23.288
2	2	11.288	26.712

Acceptance of this model over the previous model indicates the presence of a *partial association* between Variables B and C: B and C are associated even after the effects of Variable A are taken into account. Moreover, since τ_{ijk}^{abc} is *not* in the model, the nature or pattern of the *simple* associations between B and C is relatively consistent over the various levels of Variable A. Consequently, should we collapse over levels of A and examine only the $[BC]$ configuration, a marginal association between B and C will also be observed.

The distinction between marginal and partial association merits clarification. The former, a marginal association between two variables such as B and C, is the association between B and C when the third variable (Variable A) is ignored—that is, when we collapse over levels of Variable A to view exclusively the association in the $[BC]$ configuration. Realize, however, that a marginal association may be conditioned by the third variable. For example, should we deliberately adjust the association between B and C to take into account the influences exerted on this association by Variable A, and should there remain an association subsequent to this adjustment, then what remains is the partial association between B and C.

We could exercise control over the influence of the third variable by examining the *simple* associations at separate levels of the third variable. If simple associations are seen at each level of the third variable, a partial association exists. However, if simple associations are not observed at separate levels of the third variable, the partial association does not exist. Therefore, a marginal association between two qualitative variables is analogous to a zero-order correlation, when the latter is calculated between two interval measures. On the other hand, a partial association between two qualitative variables is comparable to a first-order partial correlation between interval measures.

Finally, to detect the presence of a partial association, the first-order parameter representing the association must be entertained in the last or seventh position in a seven-parameter model. For example, to determine whether there is a partial association between Variables A and B, the adequacy of fit of a six-parameter model, such as shown below, would be assessed first:

$$F_{ijk} = \tau \tau_i^a \tau_j^b \tau_k^c \tau_{ik}^{ac} \tau_{jk}^{bc}$$

Following this assessment, τ_{ij}^{ab} would be added to form a seven-parameter model. If it is determined that this seventh parameter is needed to achieve acceptable fit, then there is a partial association between A and B.

Model (8): Saturated

The saturated model is perhaps the least interesting model in that F_{ijk}'s are completely determined by observed f_{ijk}'s (see Table 5.1). Parenthetically, the number of degrees of freedom associated with a residual chi-square test of this model is zero, a number that reflects the fact that F_{ijk}'s are completely constrained. If it happens that restricted models are unable to fit observed date adequately, then attempts to advance a parsimonious explanation must be abandoned and the saturated model must be accepted and interpreted.

In general, acceptance of the saturated model indicates that partial associations are present between each pair of variables. It also indicates that the nature (pattern or direction) of *simple* associations changes in some manner when viewed over levels of the third variable. With respect specifically to 2 × 2 × 2 tables, the direction (positive vs. negative) of simple associations between any two variables is reversed over levels of the third variable. Marginal associations may or may not be present; only an examination and comparison of more restricted models will reveal the presence or absence of marginal relationships. It almost goes without saying that the interpretation and follow-up of a saturated model in the symmetrical case can be somewhat complicated. However, when interpreted within the context of an asymmetrical analysis, which we will entertain next, Model (8) presents fewer difficulties.

INTERPRETING LOGIT MODELS IN ASYMMETRICAL INQUIRY

This section deals with a special and most useful application of a log-linear analysis, called *logit-model analysis* (cf., Goodman, 1972b). Technically, a *logit* refers to a difference between two logarithms. And since we have defined parameters in log-linear models as effects, as differences between logarithms, and have found that viewing parameters as effects is most meaningful when different groups are being compared, the designation logit-model analysis is highly appropriate for the type of analysis to be discussed here.

In many respects, a logit-model analysis is similar to an ANOVA. The study of adverse impact, for example, lends itself to a logit-model analysis that can be loosely compared to a two-way ANOVA. In the study of adverse impact, the two explanatory variables have been crossed to form a 2×2 factorial arrangement where the intent is to perform an ANOVA-like analysis on Outcome at Termination of Testing, the latter being a qualitative response variable or logit variable. The majority of studies in the behavioral sciences lend themselves to asymmetrical inquiry yet, unfortunately, most current expository writings place greatest emphasis on symmetrical inquiry. An exception is the compiled writings of Goodman (1978). Goodman, in fact, has greatly advanced the use of the logit-model approach by demonstrating that distinctively different formulations and models are *not* needed to perform an asymmetrical analysis. Instead, in most cases, a logit-model analysis can be performed by generating, evaluating, and interpreting a subset of the models with which we have been working. Incidentally, this subset of models even lends itself to rudimentary forms of path analysis, a topic reserved for Chapter 7. For now, our aim is simply (a) to point out those models in Table 5.2 (or Table 5.3) that are legitimate logit models and to indicate why they are so designed, and (b) to point out the omnibus conclusion or interpretation associated with each of these models.

Legitimate Logit Models

A clue as to what constitutes a legitimate logit model can be gleaned from our past work. Examine again the models for two-dimensional tables presented in Table 4.2 and note that only Models (3) and (4) have potentially meaningful interpretations in the asymmetrical case. (As we shall see, Models (3) and (4) are logit models for a two-dimensional situation.) Now, from the context of the working example of the last chapter, it was clear that Variable A (Sex) was explanatory, whereas Variable B (Attitude) was the logit variable. It was also clear that the central question was whether females and males responded in equal proportions to each category of the attitude variable. Since we were concerned with patterns of proportional response to Variable B within Variable A, the fact that there happened to be fewer females in the sample ($f_1^a = 40$) than males ($f_2^a = 60$), and that, overall, more members of the sample opposed the amendment ($f_1^b = 70$) than supported it ($f_2^b = 30$), was not relevant to the central concern. After all, it is the investigator who determines $[A]$ (i.e., frequencies in the main-marginal of Variable A) by his or her choice of sampling scheme. And frequencies seen in $[B]$, per se, have no bearing on differences in proportional response between females and males. Hence, main-marginal differences in $[A]$ and $[B]$ are not only irrelevant, but to the extent to which they exist they should not be permitted to distort the picture of proportional response to the logit variable by sex.

For the working example, irrelevant main-marginal differences can be acknowledged and controlled by incorporating their terms (or factors) in all asymmetrical models. In other words, at very least a legitimate logit model should contain τ_i^a and τ_j^b, as does Model (3). If Model (3) should be chosen, we can conclude that there may be differences in $[A]$ and $[B]$, but beyond these differences the proportional response of females and males to levels of the logit variable is similar. But if Model (4) should prove to be the model of choice, we can say that there are differences in the proportional response of females and males to levels of the logit variable, differences that are over and above arbitrary differences in the main-marginals of A and B. In sum, to be a legitimate logit model, the model must contain all parameters that reflect potential differences not relevant to proportional response over the logit variable. The model must contain these parameters so that they can be controlled in a manner analogous to that seen in the analysis of covariance.

Let us return to a three-variable situation, a situation characterized by two explanatory variables and one logit variable. Here again, a logit model should contain all parameters that are linked to irrelevant or arbitrary differences in the main-marginals, but in addition the model must contain the first-order parameter associated with the fixed (hence, irrelevant) two-variable configuration. Consider again the study of adverse impact where it was clear that Variables *A* (Sex) and *B* (Ethnic Group) were explanatory and Variable *C* (Outcome at Termination of Testing) was the response variable. Recall that the intent of the study was to determine whether there were effects for sex, groups, or interaction (sex by groups) with respect to the outcome of testing. To pursue this intent, a sample of 100 workers was obtained, a sample in which males far outnumbered females and majority group members slightly outnumbered minority group members. These differences could have been deliberately structured or they may simply have come about as a result of sampling. In any event, we certainly do not want the uneven distributions in either [*A*] or [*B*] to influence logit-variable response. By the same token, differences in observed cell frequencies in the [*AB*] configuration are not relevant to logit response, hence they too should not be permitted to influence logit response, they too must be controlled. It follows that we must fix the [*AB*] configuration—and by doing so we can automatically fix upon both [*A*] and [*B*]—*so that we can standardize to unity the proportional response to Variable C within each cell of the* [*AB*] *configuration.* Consequently, present in any and all logit models for the working example must be the parameter estimate τ_{ij}^{ab} (or λ_{ij}^{ab}).[1] The models in Table 5.2 that qualify as legitimate logit models are Models (5), (6), (7), and the saturated model. Adhering to the practice of the last section, we will fit and interpret each of these models.

Model (5): The Completely Null Model

The elementary cell frequencies given by Model (5) have been calculated and displayed previously (see Eq. 5.6). But we know that the most meaningful view of a logit-model analysis is that of propor-

---

[1] Assuming *A* and *B* to be explanatory and *C* to be response, any model that does *not* contain λ^{ab} will not fit the observed [*AB*] configuration in the process of generating F_{ijk}'s. Instead, such a model will generate *expected* cell frequencies for [*AB*] that will most likely be at variance with observed cell frequencies in [*AB*]. And since frequencies in [*AB*] are determined by sampling, any model that does not reproduce these fixed frequencies is not a legal model for a logit analysis.

tional response over levels of the logit variable (Variable C here) from within cells of the fixed $[AB]$. Accordingly, we take the F_{ijk}'s given by this model for the cells of $[AB]$, and within these cells we standardize to unity the proportional response to levels of C. Mathematically, for the cell situated in the ith level of A and the jth level of B, the *fitted* proportion for the kth level of C is

$$P_{ijk} = F_{ijk}/f_{ij}^{ab} \tag{5.9}$$

The use of Eq. (5.9) for Model (5) on working data gives

A	B	C_1	C_2	$[AB]$
1	1	.400	.600	1.000
1	2	.400	.600	1.000
2	1	.400	.600	1.000
2	2	.400	.600	1.000

Should observed data be well fitted by this model, then no effects of any kind are present. By analogy, the outcome suggested by Model (5) is like that given by a two-way ANOVA when the F tests for both main effects and for the interaction are found to be nonsignificant. Therefore, to the extent that Model (5) fits observed data, one accepts all null hypotheses.

Model (6): Main Effects for Variable A

In addition to obligatory λ_{ij}^{ab}, Model (6) contains λ_{ik}^{ac}. To obtain patterns of proportional response to levels of the logit variable we again use Eq. (5.9), but now the F_{ijk}'s in the numerator are the F_{ijk}'s produced by, and previously tabled for, Model (6). Fitted cell proportions turn out to be

A	B	C_1	C_2	$[AB]$
1	1	.500	.500	1.000
1	2	.500	.500	1.000
2	1	.375	.625	1.000
2	2	.375	.625	1.000

Should this model fit observed data well, what ANOVA-like conclusion can be drawn? Can we not conclude first that the female and male profiles of logit-variable response are distinctly different? Can it not be said that main effects have been observed for Variable A? The fact that λ_{ik}^{ac} is needed to achieve a good fit indicates that the "difference" between C_1 and C_2 will in turn be different at A_1 and at A_2. In short, the presence here of a first-order interaction between A and C is best interpreted as the presence of main effects for Variable A. Notice, however, that profiles of response to C do not change over levels of B; hence, no main effects are indicated for levels of the ethnic group variable. Also, profiles do not differ within the cells of $[AB]$ after adjustments are made for the main effects due to A, and hence there is no evidence of interaction.

Model (7): Main Effects for Variable B

Selection of Model (7) implies that λ_{jk}^{bc} is needed to achieve reasonable fit. To the extent to which this first-order term is needed, response profiles over levels of C can be expected to vary over levels of B. That is, from an asymmetrical perspective, expect main effects for Variable B. To see whether our expectation is correct, Eq. (5.9) was used on F_{ijk}'s produced by Model (7) with the following result:

B	A	C_1	C_2	$[AB]$
1	1	.572	.428	1.000
1	2	.446	.554	1.000
2	1	.412	.588	1.000
2	2	.297	.703	1.000

It should be pointed out that the indices for Variables A and B above have been reversed so as to accommodate better a visual comparison between levels of B. And although there are concurrent main effects for Variable A, over and above these effects, it is apparent that response profiles are dissimilar over levels of Variable B.

Perhaps a clearer picture of the differences between levels of B can be obtained by collapsing over levels of A so that proportional response to levels of C can be seen first for majority workers (B_1) and then for minority workers (B_2). If this is done, 47.2% of majority workers fail to qualify for the position while 52.8% qualify. In contrast, only 31.9% of minority workers fail to qualify, and therefore 68.1% qualify. When proportional response is normalized to unity within levels of B, the main effects for B are clearly seen.

For the sake of completeness, it should be pointed out that acceptance of Model (7) per se provides no information about concurrent main effects for Variable A. These latter main effects may or may not be present: it depends on the specific situation. As we shall see, to determine whether there are meaningful differences between levels of A with respect to logit-variable response, we can either assess the magnitude of the component chi-square for Model (6) or, using a normal deviate test, perform a test of statistical significance on the estimated interaction effects associated with λ_{ik}^{ac}.

Model (8): Interaction between Variables A and B

For Model (8), the fitted proportions over levels of C for each cell in $[AB]$ are given by $P_{ijk} = f_{ijk}/f_{ij}^{ab}$. These proportions are

A	B	C_1	C_2	$[AB]$
1	1	.636	.364	1.000
1	2	.333	.667	1.000
2	1	.429	.571	1.000
2	2	.316	.684	1.000

Several interpretations are possible when the saturated model is accepted. A most straightforward interpretation is that since λ_{ijk}^{abc} is needed in the model, interaction between Variables A and B is indicated, an interpretation analogous to first-order interaction in a two-way ANOVA. (Incidentally, meaningful main effects for Variables A and B may, or may not, be present.) The consequence of observing an interaction in the study of adverse impact could be interpreted in several ways. It could be said, for example, that the pattern of response to the logit variable given by females and males who belong to the majority is distinctively different from the pattern of response observed for females and males of the minority. It can also be said that the pattern of logit-variable response observed for majority and minority applicants is different as a function of sex. In either case, interaction connotes complexity, complexity that should be acknowledged during the interpretation of results. At a later point we will have the opportunity to illustrate a substantive interpretation of a first-order interaction.

MODEL COMPARISONS AND SELECTION

The most important yet least prescribed operation in log-linear analysis is the selection of a model for subsequent interpretation. From Chapter 4 we know that the objective is to choose a model that on the one hand is parsimonious yet, on the other hand, complete enough so that it fits observed elementary cell frequencies reasonably well. We also know that choosing a model is not exclusively a statistical decision; however, both residual and component chi-square procedures can be used to great advantage when attempting to select a model.

The aim of this section is to apply and extend our knowledge of model selection to tables of three dimensions. We will begin by illustrating the use of residual and component chi-square testing on adverse impact data to compare models for acceptability-of-fit. A general strategy for model selection will then be outlined. Finally, we will attempt to select the model that appears to be most acceptable in the general symmetrical case and the most acceptable for a logit-model analysis.

Comparing Models for Acceptability-of-Fit

Residual Test Statistics. The residual chi-square is used to identify models that fit observed data so poorly that they can be eliminated immediately from further consideration. Consider the fit offered by Model (1) to the data observed in Table 5.1. Model (1) fits a value of 12.5 to each elementary cell in the $2 \times 2 \times 2$ table. Now, to assess the extent to which the frequencies fitted by Model (1) deviate from observed frequencies, a residual Pearsonian or likelihood-ratio chi-square statistic is calculated:

$$L_1^2 = 2 \sum_i \sum_j \sum_k (f_{ijk})[\ln (f_{ijk}/F_{ijk})]$$

$$= 2 \{[7[\ln (7/12.5)] + \cdots + 26 [\ln (26/12.5)]\}$$

$$= 46.94$$

Computing a residual chi-square for Model (2) yields $L_2^2 = 8.39$, which indicates that the fit provided by this model is better than that offered by Model (1). Moreover, as residual statistics are calculated successively on models in the hierarchy, their values tend to decrease monotonically until a perfect fit is realized (hence, $L^2 = 0.00$) for the saturated model. This can be seen in Table 5.4 where, for the working example, resultant residual chi-squares are displayed.

Degrees of Freedom for Residual Testing. To test for statistical significance, the number of degrees of freedom for each residual chi-square must be determined. In Chapter 4, these determinations were made intuitively. It was promised, however, that a "rule of thumb" would be offered to assist researchers with these decisions when complexities arise. The rule of thumb for complete tables, stated as an equation, is presented below:

$$df(\text{for Model X}) = \text{SBP(Sat.)} - \text{SBP(Model X)} \qquad (5.10)$$

The equation reads: the number of degrees of freedom associated with the residual testing of a given model, say Model X, is equal to SBP(Sat.), the *sum* of the number of *basic parameters* in the saturated model, minus SBP(Model X), the *sum* of *basic parameters* in Model X.

Obviously, the definition of basic parameters is key to the use of the rule. Basic parameters are parameters representing specific effects that are free to vary. When all qualitative variables are dichotomies, the number of basic parameters associated with a model is equal to the number of general parameters in that model. For example, the saturated model for a $2 \times 2 \times 2$ table has eight general parameters, and it also has eight basic parameters. But consider a saturated model for a table of size $3 \times 2 \times 3$. To count the number of basic parameters here, we begin the count with λ, the log of the geometric mean; continue to λ_i^a where we find that two of the three specific effects are free to vary since $\Sigma\lambda_i^a = 0$; continue to λ_j^b where only one of the two effects can assume any value since $\Sigma\lambda_j^b = 0$; continue to λ_k^c where two effects again are free to vary, etc. In sum, the number of basic parameters for this particular saturated model would be $1 + 2 + 1 + 2 + 2 + 4 + 2 + 4 = 18$.

As an exercise, let us use the rule to determine the ν for Model (2) of the working example. Restating Eq. (5.10) and substituting into the equation the number of basic parameters contained in Model (2) and the saturated model gives

TABLE 5.4 Adequacy-of-Fit of General Log-Linear Models for the Adverse Impact Data

Model	Residual L^2	df	p	Component L^2	df	p
(1)	46.94	7	.00*			
(2)	8.39	6	.21	38.55	1	.00*
(3)	8.03	5	.15	.36	1	.80
(4)	4.00	4	.40	4.03	1	.05
(5)	3.96	3	.27	.04	1	.90
(6)	2.94	2	.23	1.02	1	.40
(7)	.54	1	.46	2.40	1	.20
(8)	0.00	0		.54	1	.50

*$p < .0001$.

$$df(\text{Model 2}) = \text{SBP(Sat.)} - \text{SBP(Model 2)}$$
$$= 8 - 2 = 6$$

With knowledge of ν, tests of the significance of residual departure can be made routinely for each model. For our example, these tests can be found in Table 5.4.

Component Chi-squares. Notice that Table 5.4 also contains component chi-squares for models of the working example. Components, recall, are differences between residuals. The number of degrees of freedom used to test a component chi-square is also a difference—the difference between respective residual chi-squares. Finally recall that to the extent that the component is large, there is less justification for choosing a more restricted model.

A Strategy for Model Selection

Figure 5.1 is a graphical representation of the process of selecting a general model when models are arranged hierarchically. The process is as much art as it is technique. Nevertheless, in an attempt to reconstruct this process, the following three steps are offered:

Step 1. Proceeding from the bottom of the table (i.e., the

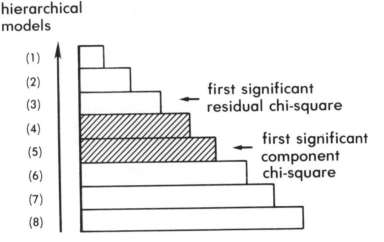

hierarchical models

Figure 5.1. Representation of a strategy for the selection of general log-linear models.

saturated model is depicted as the bottom step in the figure), examine *residual L^2*'s and continue to entertain models of increasing simplicity until the first obviously significant residual L^2 is encountered. Since the model associated with the significant residual L^2 does not fit observed data well, eliminate this model and, in general, all more restricted models from further consideration.

Step 2. Again, proceeding from the bottom of the table, but now examining *component L^2*'s, move upward until the first obviously significant component L^2 is observed. The model associated with the significant component is definitely a candidate for selection but all *less* parsimonious (less restricted) models generally can be ignored.

Step 3. If the results of Steps 1 and 2 above do *not* point unequivocally to a single model of choice, but instead identify several models that might be acceptable, then questionable parameter estimates should be tested with z tests for significance.

Before selecting a model for the working example, let us link the three steps above to Figure 5.1. As the figure shows, on the basis of the first step, Models (1), (2), and (3) have been screened out because of their significantly high residual values; these models simply do not fit observed data well. Using the second step, we are able to pass over Models (8), (7), and (6) because their component L^2's are not significant; therefore, with respect to each, we are able to proceed to a simpler model without sacrificing significant loss in ability to fit observed data. Most often when working with three-dimensional tables, the execution of these two steps will point to a single model, but not always. For example, in Figure 5.1, Models (5) and (4), represented by the shaded area, are potential choices to this point. Granted, if we move up to Model (4) from Model (5), we will experience a significant loss in ability to fit observed data, but Model (4) does not fit observed data so poorly that it yields a significant residual L^2. To feel more comfortable about the final decision, it would be helpful to know more about the fifth general parameter in Model (5), the parameter that is not present in Model (4). Hence, the third step: subject the fifth parameter, λ_{ij}^{ab}, to test. As we know, we can test the null that $H_o:\lambda_{ij}^{ab} = 0$. If this null can be rejected, a strong argument can be made for the retention of these interaction effects and the selection of Model (5). If, however, the null cannot be rejected, Model (5) could still be selected but λ_{ij}^{ab} effects would be ignored in subsequent interpretation, or a case could be made for adopting and interpreting Model (4). We will apply these principles to the working example.

Model Selection: Symmetrical Case

Assume for the sake of discussion that the adverse impact study lent itself to a symmetrical analysis. The task then would be to select a general model and to advance the symmetrical conclusion associated with that model. Using first the residual criterion, we find that we can discard Model (1) in Table 5.4 since $L_1^2(7) = 46.94$, $p < .0001$. Next, employing the component criterion, we find that we can proceed to Model (4) because the components for Models (8), (7), (6), and (5) are not significant. Thus, to this point, we have narrowed the range of possibilities to three models, Models (4), (3), and (2). Upon close study, however, Model (4) appears to be the most acceptable, particularly since the component for Model (4) indicates that a significant diminution in ability to fit data will occur if Model (3) is chosen in favor of Model (4). Since Model (4) is the model of choice, we conclude that mutual independence best describes relations among the three qualitative variables. Further, since an association appears not to be present, there is no need for additional follow-up procedures.

Factor Specification: Asymmetrical Case

Variable C is unmistakably a response variable in the study of adverse impact. The analysis, therefore, will be symmetrical and, as a consequence, we will work only with Models (5) through (8) for only these models contain λ_{ij}^{ab} and only these models account for irrelevant differences among f_{ij}^{ab}'s. Not only will we be examining fewer models, but our approach to model selection will be somewhat different. That is, rather than identifying a single model that is both parsimonious and fits observed data well, the goal in a logit-model analysis is to identify a factor (or factors) that appears to be promoting discrepancy about the null-logit model. The approach to be described is parallel in many respects to that taken when performing an ANOVA.

Consult the summary given in Table 5.5 of a particular hierarchical analysis of the adverse impact data. Notice the similarity in both appearance and function to the summary of a two-factor ANOVA. In the ANOVA, recall that an overall measure of variability about the grand mean (total sum of squares) may be partitioned into two general additive components: a component explained by the factors incorporated into the design (explained or regression sum of squares) and a residual component. Moreover, the relative size of the explained component (or the magnitude of the R^2) can be subjected to statistical test. If significance is *not* achieved, further analysis is discouraged unless there is defensible a priori argumentation to the contrary. On the other hand, if the explained composite component is found to be statistically significant, the researcher is encouraged to examine specific components for meaningfulness and significance.

Comparable features are found in analyses involving logit models. The size of the residual chi-square for the null-logit model, for example, represents the degree to which observed cell frequencies do *not* conform to those offered by the null model. In other words, the size of the residual reflects the extent to which design factors (Variable A, Variable B, etc.) are operative. Or, from a different

TABLE 5.5. Summary Table of the Logit-Model Analysis of Adverse Impact Data

Model/Source	L^2	df	p
(5) Null/Total	3.96	3	.27
(6) Due to Sex	1.02	1	.40
(7) Due to Ethnic Group, Given Sex	2.40	1	.20
(8) Due to Interaction	.54	1	.50

Note. The likelihood ratio L^2 for Model (5) is a residual chi-square; other L^2's are component chi-squares.

perspective, the residual chi-square for Model (5) can be viewed as a component chi-square defined as $L^2_{5-8}(3) = 3.96 - 0.00 = 3.96$. From this perspective, the composite influence of specific effects, analogous to sum of squares regression in the ANOVA, can be examined and tested for statistical significance. Usually, this composite test is considered first in a logit model analysis.

Taken together, are the main and interaction effects specified in Table 5.5 strong enough to render Model (5) a poor fit? Since $L^2_5(3) = 3.96$, and since this residual is nonsignificant ($p < .26$), the answer is *no*. Model (5), by statistical criteria alone, cannot be rejected as a viable explanation for observed data. Be alert to the fact, however, that the test in question is a composite test and composite tests tend to be more conservative than tests performed on constituent components. It is possible, therefore, for the composite test to fail to achieve significance yet significance could be achieved for tests performed on specific components. Should this occur, it is generally advisable to ignore the significant results associated with specific effects. After all, if the overall test is not significant, collectively tests on specific components carry with them a higher Type I error rate than is nominal. Even so, an exception is frequently seen when the results of component tests are consistent with strong a priori predictions. That is, if, on the basis of theory, predictions made in advance of data analysis happen to be supported by the results of specific tests, then despite the outcome of the composite test the investigator is justified in pursuing specific results. In any event, follow-up procedures constitute a moot issue for the outcome summarized in Table 5.5, for all tests failed to demonstrate statistical significance. We must conclude, therefore, that evidence is not at hand to support the contention that the testing program is having an adverse impact on minorities and females.

A NOTE ON THE ADDITIVITY OF COMPONENT L^2's

Before considering another example, let us take a harder look at the anatomy of component chi-squares. To begin, let us consider the discrepancy between a situation where all elementary cells have the same frequencies and the situation that we observed in Table 5.1 for the adverse impact study. Think of this discrepancy as a "grand component," a component that reflects the empirical world's inability to fit the simplest of all explanations, that given by Model (1). This so-called grand component is

$$L^2_{1-8} = 46.94 - 0.00 = 46.94$$

which we recognize as the residual chi-square for Model (1). This residual can be thought of as a measure of "poorness-of-fit," a measure of the extent to which the actual data of Model (8) do not conform to the simple state of affairs offered by Model (1).

Now, rhetorically, we ask: What sources (variables and/or variables in combination) seem to be responsible for the poorness-of-fit, the magnitude of which is 46.94 residual units? From an examination of Table 5.4, it can be inferred that the uneven distributions in [A], [B], and [C] promote most of the observed deviation away from the expectations of Model (1). In fact, taken together, main-marginals are responsible for 42.94 of residual units of poorness-of-fit as shown below:

$$L^2_{1-4} = 46.94 - 4.00 = 42.94$$

Similarly, the magnitude of the poorness-of-fit about Model (1) that can be attributed to first-order interactions is

$$L^2_{4-7} = 4.00 - .54 = 3.46$$

Finally, that attributable to the second-order interaction is

$$L^2_{7-8} = .54 - 0.00 = .54$$

Remember, however, that main-marginal differences are not meaningful in a logit-model analysis. Also, assuming Variable C to be the logit variable, first-order effects associated with λ^{ab}_{ij} are also not relevant considerations. Thus, even though the collective first-order component (the magnitude of which was found to be 3.46 residual units) can be broken out into three specific first-order components, only the components associated with λ^{ac}_{ik} and λ^{bc}_{jk} are relevant, for we are no longer concerned with residual deviation about the null condition posited by Model (1), but rather with residual deviation about a null condition posited by Model (5).

Perhaps a number of these points can be clarified by the following summary of the outcomes of tests of significance on specific components for the three first-order interaction models:

Model (5). The test on the component associated with the fifth general parameter is a test on the null that $H_o: \lambda^{ab}_{ij} = 0$. Should this null be rejected, in the symmetrical case, we would conclude that a significant marginal association is present between Variables A and B. In our logit-model example, however, this null hypothesis has no meaning. Marked λ_{ij}^{ab} first-order effects may or may not be present; it depends on the sampling scheme used.

Model (6). The test on the component associated with the sixth parameter is a test of the null $H_o: \lambda_{ik}^{ac} = 0$. In the symmetrical, a significant result indicates the existence of a marginal association between A and C. For our logit-model analysis, a significant result means that there are differences between females and males with respect to how they respond to the logit variable.

Model (7). The test of the component for the full first-order model is a test on $H_o: \lambda_{jk}^{bc} = 0$. In the symmetrical, if significant, a partial association between B and C is indicated. Extending this notion to a logit-model analysis, significance indicates that members of the two ethnic groups differ in their response to the logit variable and, more importantly, *these differences are present subsequent to partialling out* (or adjusting for) *any differences in logit-variable response that might exist between levels of Variable A.*

From a symmetrical perspective, therefore, the collective first-order component (i.e., $L_{4-7}^2 = 3.46$) can be viewed as the sum of two components linked to marginal associations and a component reflecting the strength of a partial association. Understand that the order in which respective parameters appear in the model determines whether its associated component represents a marginal or a partial association. For example, to test for a partial association between Variables A and B, λ_{ij}^{ab} would be entered last in a seven-parameter model, and the component for this partial association would be the difference in residuals between the model in question and a six-parameter model—a model identical to the seven-parameter model except for the λ_{ij}^{ab} term.

But from an asymmetrical perspective, even though the collective first-order component can be broken out into three components, we are interested in only two of these. To be specific, for the working example interest is focused on the following component:

$$L_{5-7}^2 = 3.96 - .54 = 3.42$$

since this component reflects the magnitude of the discrepancy between observed cell frequencies and those posited by Model (5), the completely null logit model.

There are several ways to account for L_{5-7}^2 in a logit-model analysis. One is to enter λ_{ik}^{ac} as the sixth parameter in a six-parameter

TABLE 5.6. A Comparison of Three Approaches to the Calculation, Interpretation, and Reporting of Effects in a Logit-Model Analysis

Model Number	Order in which Parameters Are Entered					
	AB, AC, BC, ABC		*AB, BC, AC, ABC*		Both Orders	
	Source	L^2	Source	L^2	Source	L^2
(5)		3.96		3.96		3.96
(6)	*A*	1.02	*B*	2.43	*A|B*	.99
(7)	*B|A*	2.40	*A|B*	.99	*B|A*	2.40
(8)	*AB*	.54	*AB*	.54	*AB*	.54

Note. The sum of component chi-squares for the conservative approach, where partialled main-effects are reported, does not equal the residual for the null-logit model.

model, then structure a seven-parameter model where λ_{jk}^{bc} is the seventh parameter, and then proceed as before. Another way would be to reverse the order of entering these two parameters. That is, enter λ_{jk}^{bc} in as the sixth parameter in a six-parameter model and λ_{ik}^{ac} as the last parameter in a seven-parameter model. Doing this would permit a test of main effects for Variable *B* (Ethnic Groups) and a test for the main effects of Variable *A* (Sex) *subsequent to partialling out main effects for B.* There is still a third alternative, a very conservative alternative that is analogous to an approach often taken in the ANOVA when cell frequencies are not equal, which is to calculate, interpret, and report both main effects for *A* and *B* subsequent to partialling out the effects of the other variable. The three approaches as applied to the adverse impact data are presented in Table 5.6.

THE EFFECTS OF REFLECTIVE TEACHING: A DIDACTIC EXAMPLE

In this section we describe an educational experiment, the Reflective Teaching Study, reported by Holton and Nott (1980). Aside from this writer's desire to expose readers to a piece of experimental research—so as to assuage the impression that log-linear methods are only used for descriptive research—another reason for choosing this particular study is to show how these researchers handled a potentially troublesome pre-/posttest variable to avoid violating the assumption of response independence. Also, as will become apparent, the study is unequivocally asymmetrical and is therefore representative of the majority of studies in behavioral work. We will, however, approach the research on reflective teaching from both symmetric and asymmetric perspectives to reinforce the widest possible number of understandings. Finally, we will introduce several conventions and formats that are used by the popular and widely available BMDP/4F computer program to report log-linear results.

An Overview of the Reflective Teaching Experiment

The 101 subjects in this experiment were teachers in training enrolled in a large midwestern university. Subjects were enrolled in four intact classes two of which were assigned at random to be exposed to *reflective teaching*, a new form of on-campus laboratory teaching experience, while the remaining two classes served as controls. One hypothesis was that exposure to reflective teaching would enhance the ability of the students to express themselves in an analytical manner when discussing the process of learning.

Data of immediate interest were the written responses of students to the stimulus "When I think about learning" gathered prior to and following the experiment. Pretest responses to the stimulus sentence were subjected to a grammatical structural analysis, which ultimately led to each student being placed into one of four categories based on his or her predominant mode of response. The grammatical categories were: (a) analytical, (b) evidential, (c) declarative, and (d) indeterminate, the latter containing subjects whom judges could not reliably classify into one of the three other categories. Identical methods were used with respect to the posttest responses provided by subjects. There were, therefore, three qualitative variables, namely:

Variable A, a polytomy, was the *pretest* response variable. On the basis of their judged predominant mode of response, subjects were assigned to one of the four categories mentioned above. This variable was an explanatory variable in the logit-model analysis.

Variable B, a dichotomy, was the *treatment* variable; it consisted of 55 subjects from the experimental group who were exposed to reflective teaching and 46 students who were members of the control groups. This variable, obviously, was the principal explanatory variable in the experiment.

Variable C, was the *posttest* variable, and in this experiment it was the response variable. As noted, subjects were classified into one of four grammatical categories.

Before we look at the data, consider the way that these researchers handled pre-/posttest response. At first sight, it might seem desirable to combine these two sets of data to form a repeated measurements variable as is often done in classical experimental design work. Such a combination would not constitute good practice here, however. After all, treating pre-/posttest response as a repeated measure would result in each subject appearing twice in the contingency table, which would probably lead to a violation of the basic assumption that tabulations are independently determined. Instead, pretest and posttest responses were treated as separate variables, variables that were crossed with the treatment variable to form a $4 \times 2 \times 4$ contingency table in which subjects were placed into one—and only one—of the 32 elementary cells. It must be admitted, however, that these researchers were fortunate in that the correlation between pretest and posttest response was not so strong as to result in an excessive number of empty cells in the $[AC]$ configuration.

Observed tabulations for the study are shown in Table 5.7. Note that there is a void cell, a zero, in the table. The discussion of void cells will be presented in Chapter 7; for now we simply state that the presence of void cells does not preclude a straightforward

TABLE 5.7. Cross-Classification of Subjects in the Reflective Teaching Experiment by Pretest Response, Treatment Groups, and Posttest Response

Pretest Response	Treatment Group	Posttest Categories			
		C_1	C_2	C_3	C_4
A_1	B_1	8	2	1	4
A_1	B_2	1	2	2	4
A_2	B_1	6	4	2	5
A_2	B_2	1	14	5	3
A_3	B_1	1	2	2	1
A_3	B_2	1	0	1	3
A_4	B_1	8	4	1	4
A_4	B_2	2	4	1	2

log-linear analysis if these cells are a result of sampling artifacts (i.e., sampling zeros)—as opposed to structural artifacts (i.e., structural zeros, e.g., pregnant males)—and if the void cells are not overly numerous. Both requisites are satisfied in the working example. Incidentally, to proceed mathematically a small numerical quantity, usually .5, is added to each cell count to facilitate the computation of chi-square statistics.

Screening Procedures

For a table of given dimensionality, there are numerous ways that models can be arranged hierarchically. Moreover, within a given hierarchical arrangement, there are numerous distinct models that can be structured. Thus, in the search for the most acceptable model, the number of potential arrangements and the number of distinct models can pose a problem—especially if the table is large and there is little a priori theory to guide the search.

Before we consider large tables, consider for a moment a simple two-way table and the number of models that are possible for that table. In Chapter 4, four distinct models were presented in a particular hierarchical arrangement (see Table 4.1). Recall that we began with the one-parameter model and successively added terms in alphabetical order until the four-parameter saturated model was created. A different arrangement, however, could have been presented. We could have unfolded an arrangement by adding τ_j^b to τ to form the second model, τ_i^a to τ and τ_j^b to form the third model, etc. In short, for two-way tables, there are two possible hierarchical arrangements that together encompass seven distinct models. Admittedly, problems of model specification and choice are relatively minor when dealing with two-way tables. But as the number of variables increases, unfortunately so do the number of potential choices. For three-way tables, for example, there are not two different hierarchical arrangements; there are in fact 72 different arrangements. These 72 arrangements embrace a total of 142 distinct models. For four-way tables, the number of possible arrangements and distinct models is even greater.

Clearly, as has been pointed out by Brown (1976), ". . . some screening of the effects or interactions in the log-linear model is necessary in order to limit the number of models whose tests-of-fit need to be evaluated" (p. 38). Essentially, Brown has proposed two screening procedures that can help to reduce the number of possible models and conserve effort, time, and money. The first is to *evaluate full models for goodness-of-fit* so that choice can be narrowed to a "family" of specific models. The second screening procedure, useful when the most acceptable model appears to contain more than four parameters, is to *evaluate the marginal and partial associations* linked to interaction terms for the purpose of determining which terms are needed and which terms can be deleted from the model. It is most fortunate that these two screening procedures have been incorporated as options in the BMDP/4F computer program. Let us illustrate these procedures using the Holton-Nott data as analyzed by BMDP/4F.

Full Model Goodness-of-Fit. To effect economy, models in hierarchy can be grouped into sets according to the presence of like-order terms. These sets will be called families. Consider, for example, a typical hierarchical arrangement for a three-way table where models can be grouped into four families. Obviously, the one-parameter model constitutes a special case, a family containing only one model. Models that, in addition to the geometric mean parameter, also contain main-marginal parameters comprise another family. In this family, Model (4) is designated the full model because it contains all three main-marginal effects. Next, Models (5), (6), and (7) contain at least one first-order interaction term, and thus they belong to the family of first-order models where the most complete model, Model (7), is the full first-order model. Finally, there is the saturated model, unique in that it alone has a second-order term. Observe that we have reduced a number of specific models into a fewer number of "full" models, a reduction that is more pronounced in arrangements for tables of four and five dimensions. One aspect of screening is the performance of conventional goodness-of-fit analysis on just the full models. Often, the results of screening will point to a family of models where acceptability will be found.

Table 5.8 is *the first screening table*, a table that essentially appears upon request in BMDP/4F output and which, in this instance, summarizes the results of residual chi-square tests of full models for the Holton-Nott study. (The "K–FACTOR" column in this table is the convention used in BMDP/4F to identify full models.) The table is approached as before. That is, simple models that fit observed data poorly are identified by their large residual chi-squares and immediately excluded from serious consideration. In Table 5.8 it is apparent that the one-parameter model can be excluded from further consideration because of its relatively large and statistically significant chi-square.

TABLE 5.8. Residual Chi-square Screening Table Applicable to the Holton-Nott Data

Full Model	K FACTOR*	D.F.	LR CHISQ	PROB
(1)	0	31	54.45	.0057
(4)	1	24	32.96	.1049
(7)	2	9	8.09	.5252

Note. Results suggest the elimination of Model (1).

*The K FACTOR refers to the convention used in BMDP/3F and BMDP/4F to denote full models. When K = 1, for example, reference is made to the full main-marginal model, Model (4).

Table 5.9 is *the second screening table.* This table summarizes the results of tests on component chi-squares. As before, the objective is to see whether relatively complete models can be passed over in favor of models containing fewer terms. Since the component for the saturated model is nonsignificant, $L^2_{7-8}(9) = 8.09$, $p < .53$, we proceed to the full first-order model. Here, the component approaches significance at the .05 level: $L^2_{4-7}(15) = 24.87$, $p < .052$. Since we are presently engaged in a screening process and not making a final decision, the component should be deemed practically significant. In other words, it would *not* be sagacious at this point to pass over all models in the first-order family in favor of entertaining Model (4).

We pause to summarize our thinking to this point. On the basis of residual testing (Table 5.8) we were able to eliminate from potential choice Model (1), and on the basis of component testing (Table 5.9), we have excluded the saturated model. Thus, all things considered, the model of ultimate choice will likely be in the family of first-order

TABLE 5.9. Component Chi-square Screening Table Applicable to Holton-Nott Data

Models	K FACTOR	D.F.	LR CHISQ	PROB.
(1)-(4)	1	7	21.49	.0031
(4)-(7)	2	15	24.87	.0517
(7)-(8)	3	9	8.09	.5252

Note. Results suggest that a model more parsimonious than the saturated model will be acceptable.

models. If further attempts to screen models were to terminate at this point, the fit of at least the three first-order models and Model (4) would need to be examined. Details relating to the fit of the latter model would be desirable because it represents the maximum degree of parsimony that might be achieved. Remember that the residual fit of Model (4) was not so poor as to issue a significant residual test statistic. Screening need not terminate here, however.

Assessing Marginal and Partial Associations. If residual and component screening procedures point to a family of models that contain interaction terms, such as occurred in the Holton-Nott study, then an additional screening procedure proposed by Brown can be used to judge the relative importance of specific interaction terms. Unfortunately, decisions concerning whether a first-order term should or should not be in a model are complicated by the fact that these terms are not independent of one another, they are nonorthogonal, and as a result there is no single test that can be performed to determine the importance of a specific interaction term. Recall the complexities encountered in the last section when we learned that due to non-orthogonality, the collective first-order chi-square component (i.e., L^2_{4-7}) could not be partitioned into three additive components such that each component was reflective of a two-variable association independent of the influence of the third variable. Of importance here is that when trying to decide if an interaction term is important enough to be retained in a model, a statistical test of both the marginal and partial association corresponding to the term would be most helpful.

Table 5.10 presents the desired ad hoc tests of partial and marginal association performed on the Holton-Nott data. This tabular information is available upon request on the BMDP/4F program. To use the table, one examines each effect and, as Brown

TABLE 5.10. Partial and Marginal Associations Computed on the Holton-Nott Data

		PARTIAL		MARGINAL	
EFFECT	D.F.	LR CHISQ	PROB	LR CHISQ	PROB
AB	3	1.51	.6807	3.65	.3019
AC	9	8.77	.4585	10.92	.2815
BC	3	10.31	.0161	12.45	.0060

suggests, if both the partial and marginal test statistics are sufficiently large, then that effect is definitely needed in subsequent models. On the other hand, if both test statistics are small, the term can be ignored when constructing ensuing specific models. An exception, of course, is made when a logit-model analysis is being performed and the term in question represents effects fixed by the sampling plan (e.g., λ_{ij}^{ab} is the fixed term needed in all logit models for the Holton-Nott study). If a term cannot be included or excluded because test results are not uniformly strong or weak, then the term should be retained and a final decision made when specific models are fitted to observed data.

Let us examine Table 5.10. If we were attempting to identify a *general* model for an asymmetrical analysis, only λ_{jk}^{bc} need definitely be retained. Hence, for the symmetrical case, the screening information provided by Tables 5.8, 5.9, and 5.10 suggests that one of the following two models will likely be judged most acceptable:

Fitted Marginals	Log-Linear Models
$[A]$, $[B]$, $[C]$	$F_{ijk} = \tau \tau^a \tau^b \tau^c$
$[BC]$, $[A]$	$F_{ijk} = \tau \tau^a \tau^b \tau^c \tau^{bc}$

But the Holton-Nott study is not symmetrical inquiry; it is clearly asymmetrical. Therefore, the table summarizing their logit-model analysis will present Models (5) through (8). From the information given in Table 5.10, we anticipate that there will be no significant main effects for Variable A (Pretests) but significance will be observed for the main effects of Variable B (Treatments).

Adequacy-of-Fit of General Models

Even though the Holton-Nott study was not symmetrical, to gain practice in selecting general models assume for the moment that it was. Also, for the moment, put aside all information gleaned from our discussion of screening procedures so that, in a sense, we can start from scratch. The task is to examine the general models and associated goodness-of-fit data presented in Table 5.11 for the purpose of identifying the most acceptable model.

The strategy for model selection outlined earlier in this chapter suggests that residual chi-squares be examined first to eliminate models that obviously do not fit observed data well. Using the residual criterion on the models of Table 5.11 reveals that Model (1)

TABLE 5.11. Adequacy-of-Fit of General Hierarchical Models to Holton-Nott Data

Model	Marginals Fitted	Residual			Component		
		L^2	df	p	L^2	df	p
(1)	n	54.45	31	.006			
(2)	A.	39.48	28	.074	14.97	3	.002
(3)	A, B.	38.79	27	.066	.69	1	.405
(4)	A, B, C.	32.96	24	.105	5.83	3	.120
(5)	AB, C.	29.31	21	.107	3.65	3	.302
(6)	AB, AC.	18.40	12	.104	10.91	9	.282
(7)	AB, AC, BC.	8.09	9	.525	10.31	3	.016
(8)	ABC.	0.00	0	1.000	8.09	9	.525

can be dismissed from serious consideration. Moreover, Models (2) and (3) do not appear to hold much promise even though their residual chi-squares technically are not statistically significant at the .05 level. Next, beginning at the bottom of the table and using the component criterion, we encounter the first significant component at Model (7). Consequently, the saturated model can be passed over. But is it judicious to abandon Model (7) in favor of parsimony and Model (6)? The component $L^2_{6-7}(3) = 10.31$, $p < .016$ suggests a significant increase in residual chi-square (hence, increase in "poorness-of-fit") should Model (6) be chosen. Further, if Model (6) were to be adopted, it would mean the exclusion of λ^{bc}_{jk}, yet we know from Table 5.10 that this was the only first-order term that manifested a statistically significant partial and marginal association. Unquestionably, of the eight models presented in the table, Model (7) is the model of choice.

But, Model (7) per se is not yet totally acceptable. From Table 5.11 it can be seen that the components for both Models (6) and (5) are, relatively speaking, small and statistically nonsignificant. This means that the marginal associations between Variables A and C and between Variables A and B are not statistically significant. (Also, we know from Table 5.10 that the partial associations between these pairs of variables are not significant.) Thus, if the law of parsimony is to be served, λ^{ac}_{ik} and λ^{ab}_{ij} should be dropped from Model (7), giving us the more restricted model:

$$\ln F_{ijk} = \lambda + \lambda^a + \lambda^b + \lambda^c + \lambda^{bc}$$

The model above is clearly the model of choice if the mode of inquiry is symmetrical. Its acceptance means that there is a significant association between Variables B and C and that associations between the two remaining pairs cannot be documented. Although not always of major interest in a symmetrical analysis, should estimates of the values of parameters be desired the researcher would fit the above model (by fitting the $[BC]$ and $[A]$ marginals) and, using the F_{ijk}'s generated by this model, calculate the effects of interest.

Adequacy-of-Fit of Logit Models

A summary of the logit-model analysis reported by Holton and Nott is contained in Table 5.12. We first look at the table to see if the residual chi-square for the null-logit model is statistically significant. At the .05 level, it is not: $L_5^2(21) = 29.32$, $p < .11$. Thus, it would seem that the results of this composite test would discourage rejection of the null logit as an explanation for observed frequencies. But since Holton and Nott posited strong predictions in advance of the data analysis, they felt justified in examining tests on specific effects. Accordingly, the component chi-square for Model (7) was noted to be significant, i.e., $L_{6-7}^2(3) = 10.31$, $p < .016$. Furthermore, since λ_{jk}^{bc} was entered last in this full first-order model, the significant component here meant that differences exist between treatment groups with respect to how they responded to the logit variable (Variable C) even after pretest differences in logit response have been taken into account. The situation is analogous to an analysis of covariance (ANCOVA) where posttest differences between treatment groups are evident after adjustments are made for initial pretest differences.

TABLE 5.12. Summary of Logit-Model Analysis Reported in the Holton-Nott Study

Model/Source	Marginals Fitted	L^2	df	p
(5) Null/Total	AB, C.	29.32	21	.106
(6) Due to Pretest	AB, AC.	10.92	9	.282
(7) Due to Treatment Given Pretest	AB, AC, BC.	10.31	3	.016
(8) Due to Interaction	ABC.	8.09	9	.525

Following-up the Logit-Model Analysis

To this point, all that can be said is that experimental group members did not respond to the logit variable in the same way as did members of the control group. But the logit variable was a polytomous variable. It consisted of four categories of posttest grammatical response: (a) analytical, (b) evidential, (c) declarative, and (d) indeterminate. Hence, we do not know whether experimental and control subjects responded differently in one, two, three, or all four posttest categories; nor do we know the direction of the difference or differences. One of the many salutary features of log-linear models, however, is that contained within their structure are indicators of the strength and directionality of differential response. These built-in indicators are the effect parameters (lambdas or taus), and of immediate interest are those effects associated with the λ_{jk}^{bc}'s, the estimated first-order effects between the treatment and the posttest variable, effects which are to be viewed as main effects in this logit-model analysis.

But what model should be fitted to provide the numerical values of the λ_{jk}^{bc}'s? This is not a trivial question, for the respective values of λ_{jk}^{bc}'s are model-dependent; they will increase or decrease in value as additional terms are added to the model. In fact, the selection of a follow-up model can be as much the subject of debate as the selection of a general model. Although there are few definitive rules of conduct in this work, it is, nevertheless, recommended that at very least the effects of interest (the λ^{bc}'s) be examined in two different models: the most parsimonious model that constitutes an acceptable fit and, if the table is not of extremely high dimensionality, the saturated model. The former will generally prove to be most useful and informative because the effects that appear in this model will be relatively "pure." That is, the effects of interest in the most acceptable logit model will reflect the state of affairs given that the most acceptable model happened to fit observed data perfectly. On the other hand, the effects of interest in the saturated model will reflect the "contaminating" influences of additional sources. Should there be *serious* discrepancies between the effects of interest in these two models, it is likely that an important term or terms has been omitted from the logit model deemed acceptable.

In the current example, since λ^{ac} and λ^{abc} were found to be inconsequential, the most restricted logit model that appears to provide a reasonable fit is one that fits marginals $[AB]$ and $[BC]$. The saturated model, of course, is Model (8) in Table 5.11. The suggestion is that both models should be fitted so that the λ_{jk}^{bc}'s in each can be studied to determine the nature of the main effects for levels of the treatment variable. These two sets of effects are displayed in Table 5.13.

In the table, the lambda effects belonging to the most acceptable logit model suggest that differential response by experimental and control members to the posttest variable was least pronounced in the indeterminate category and most pronounced in the analytical category. Moreover, the algebraic signs of the lambdas for the latter category suggest that a higher proportion of experimental subjects had their remarks judged as analytical than was the case for control subjects. Notice too that the λ^{bc}'s belonging to the saturated model tend to reinforce the patterns of differential response that have just been mentioned.

So we see that estimates of lambda or tau effects can help us discern the magnitude of group difference and are most useful in determining the direction of difference. By themselves, however, they do not tell us if suspected differences are statistically significant. To establish significance, estimates of effects must be subjected to test in a manner similar to that seen in the last chapter. z tests were performed on null hypotheses that λ^{bc}'s are equal to zero; the results are reported in Table 5.13. Although a total of eight z statistics are reported for each of the two models, realize that only three tests, in effect, are being performed, for in each 2×4 table the results of five tests will be determined by the outcomes of three basic tests.

Essentially, the z tests reported in Table 5.13 are of the same form as noted in the previous chapter. That is

$$z(\lambda_{jk}) = \lambda_{jk}/\text{S.E.}(\lambda_{jk})$$

The computation of the standard error (S.E.), however, is not as straightforward as it was before because we are no longer dealing with a simple fourfold table. If we were, then Eq. (4.27) would still suffice. In fact, as long as we are dealing exclusively with dichotomous variables, then a simple extension of Eq. (4.27) can be used with tables of higher dimensionality to obtain the S.E. of *all* estimated effects in the saturated model. For example, for a $2 \times 2 \times 2$ table, where there are a total of seven *basic* effect parameters (but eight basic parameters), the S.E. common to all basic effects is

$$\text{S.E.}(\lambda) = [\Sigma(1/f_{ijk})]^{1/2}/(abc) \tag{5.11}$$

where the summation is understood to be over all elementary cells

TABLE 5.13. Observed f^{ac}'s, Lambdas, and Tests on Lambdas for Two Models Fitted to the Holton-Nott Data

| Posttest Variable | Acceptable Logit Model | | | | | | The Saturated Model | | | |
| | | Experimental | | | Control | | Experimental | | Control | |
	f^{bc}	λ^{bc}	z	f^{bc}	λ^{bc}	z	λ^{bc}	z	λ^{bc}	z
Analytical	23	.557	3.07	5	−.557	−3.07	.420	2.08	−.420	−2.08
Evidential	12	−.306	−1.94	20	.306	1.94	−.078	−.37	.078	.37
Declarative	6	−.239	−1.24	9	.239	1.24	−.232	−1.09	.232	1.09
Indeterminate	14	−.013	−0.08	12	.013	0.08	−.109	−.60	.109	.60
Totals	55	0.000		46	0.000		0.000		0.000	

150 /

of the table. Goodman (1970) and Bishop et al. (1975), among others, have used Eq. (5.11) to calculate the S.E. of lambdas for the saturated model of tables defined by dichotomies and then have used this S.E. to assess the significance of lambdas appearing in unsaturated models. This procedure can be defended on the grounds that the S.E. computed for all λ's on the saturated model represents a conservative estimate of exact standard errors for more restricted models. A more versatile method that often yields exact standard errors for unsaturated models has been proposed by Lee (1977) and is used to calculate S.E.'s and z tests in the BMDP/4F program.

Returning to the research of Holton and Nott, results of significant testing revealed that effects associated with the analytical category were statistically significant in the follow-up model as well as in the more turbid saturated model. Holton and Nott, therefore, appropriately confined their conclusions to differences between treated groups with respect to posttest analytical response, conclusions that were consistent with their a priori hypothesis.

6

Tables of High Dimensionality

The versatility and power of log-linear analysis is realized most fully when contingency tables are of high dimensionality. It is hoped that this chapter, which is designed to show how the principles and procedures of previous chapters can be applied to tables of four and five dimensions, will provide readers with the experience and confidence needed to tackle the analysis of even larger tables. This chapter opens with an overview of selected features and principles as seen and applied in the analysis of four-dimensional tables. The middle and latter sections of this chapter are devoted to two didactic illustrations where emphasis is given to asymmetrical inquiry.

FOUR-DIMENSIONAL TABLES

Layout and Notation

To analyze tables defined by four crossed qualitative variables, the system of notation established in Chapter 3 must be extended to accommodate the new variable, the fourth variable. Adhering to earlier practice, the first three variables and their respective subscripts remain A ($k = 1, 2, \ldots, a$), B ($j = 1, 2, \ldots, b$), and C ($k = 1, 2, \ldots, c$). The fourth variable will be denoted D, with subscript m where $m = 1, 2, \ldots, d$.[1] An unspecified *observed* elemen-

---

[1] Lowercase "m" is preferred over the lowercase letter "l" as the subscript for Variable D because the appearance of "l" too closely resembles the numeral 1 providing, therefore, too many opportunities for confusion.

tary cell frequency will be f_{ijkm}. Accordingly F_{ijkm} denotes an *expected* elementary frequency.

For tables defined by four variables, there are six distinct two-variable (first-order) configurations, four three-variable (second-order) configurations, and, of course, the observed four-variable configuration [*ABCD*]. Finally, realize that *n* subjects are cross-classified on the basis of the four variables, respecting the requisite conditions discussed in Chapters 3 and 4, so that each subject belongs to one (and only one) of the elementary cells in the $a \times b \times c \times d$ contingency table.

General Models in a Hierarchical Arrangement

When models are arrayed hierarchically by ascending complexity as before, we have: (a) a one-parameter completely null model, (b) four main-marginal models, (c) six first-order models, (d) four second-order models, and (e) a sixteen-parameter saturated model. In its multiplicative form, the saturated model is:

$$F_{ijkm} = \tau\tau_i^a\tau_j^b\tau_k^c\tau_m^d\tau_{ij}^{ab}\tau_{ik}^{ac}\tau_{im}^{ad}\tau_{jk}^{bc}\tau_{jm}^{bd}\tau_{km}^{cd}$$
$$\times\ \tau_{ijk}^{abc}\tau_{ijm}^{abd}\tau_{ikm}^{acd}\tau_{jkm}^{bcd}\tau_{ijkm}^{abcd} \tag{6.1}$$

Due to the size of models with which we will be working, some abbreviation and even abridgement will be needed in ensuing presentations. Table 6.1, for example, attempts to display a particular hierarchical arrangement of models by using only superscripts to denote all but the first parameter in each model.

Screening Models for Acceptability of Fit

When inquiry is symmetrical and theory is weak, the task of narrowing the range of models to a promising few can be both time consuming and costly. Just consider the effort required to fit and evaluate the 16 models presented in Table 6.1—never mind the effort that would be required to evaluate individually the hundreds of distinct models not explicitly presented in Table 6.1. Without doubt, in the absence of a priori interest in specific models, screening procedures such as those discussed in the last chapter are needed. If used intelligently, these procedures could greatly reduce the number of computer runs needed to choose an acceptable model. Therefore, for practice, let us extend the screening procedures introduced in the last chapter to the more challenging four-dimensional situation.

TABLE 6.1. A Hierarchical Arrangement of Models for Tables of Four Dimensions

Model No.	Log-Linear Models	Marginals Fitted
(1)	λ	n
(2)	$\lambda + a$	$A.$
(3)	$\lambda + a + b$	$A, B.$
(4)	$\lambda + a + b + c$	$A, B, C.$
(5)	$\lambda + a + b + c + d$	$A, B, C, D.$
(6)	$\lambda + a + b + c + d + ab$	$AB, C, D.$
(7)	$\lambda + a + b + c + d + ab + ac$	$AB, AC, D.$
(8)	$\lambda + a + b + c + d + ab + ac + ad$	$AB, AC, AD.$
(9)	$\lambda + a + b + c + d + ab + ac + ad + bc$	$AB, AC, AD, BC.$
(10)	$\lambda + a + b + c + d + ab + ac + ad + bc + bd$	$AB, AC, AD, BC, BD.$
(11)	$\lambda + a + b + c + d + ab + ac + ad + bc + bd + cd$	$AB, AC, AD, BC, BD, CD.$
(12)	$\lambda + a + b + c + d + ab + ac + ad + bc + bd + cd + abc$	$ABC, AD, BD, CD.$
(13)	$\lambda + a + b + c + d + ab + ac + ad + bc + bd + cd + abc + abd$	$ABC, ABD, CD.$
(14)	$\lambda + a + b + c + d + ab + ac + ad + bc + bd + cd + abc + abd + acd$	$ABC, ABD, ACD.$
(15)	$\lambda + a + b + c + d + ab + ac + ad + bc + bd + cd + abc + abd + acd + bcd$	$ABC, ABD, ACD, BCD.$
(16)	$\lambda + a + b + c + d + ab + ac + ad + bc + bd + cd + abc + abd + acd + bcd + abcd$	$ABCD.$

Note. To promote economy of expression, parameter estimates in log-linear models are designated by their respective superscripts—with the exception of λ, the "geometric mean" parameter.

TABLE 6.2. Screening Tests on Full Models Applied to the McLean Data

Model No.	K FACTOR	\multicolumn{3}{c}{Residual}	\multicolumn{3}{c}{Component}
		D.F.	LR CHISQ	PROB	D.F.	LR CHISQ	PROB
(1)	0	15	929.64	.000			
(5)	1	11	829.83	.000	4	99.81	.000
(11)	2	5	10.06	.074	6	819.77	.000
(15)	3	1	2.53	.112	4	7.53	.110
(16)	4	0	0.00		1	2.53	.112

Note. The models of this table correspond by number to the models displayed in Table 6.1.

The data to be shown were obtained by McLean (1980) in an unpublished doctoral study that will be discussed in greater detail in the next section. For now, we need only know that each of the four variables in the McLean study was a dichotomy, the size of the sample was very respectable ($n = 896$), and that the 3F program in the BMDP package was used to print out screening tables of residual and component chi-squares similar to Tables 5.8 and 5.9.[2] In this chapter, residual and component chi-squares for full models of the McLean study are combined into one table, namely Table 6.2.

Examining residual chi-squares, from the bottom to the top of the table, reveals that the first significant chi-square is that associated with Model (5), i.e., $L_5^2(11) = 829.83$, $p < .000$. Therefore, the family of main-marginal models and the completely null model can be excluded from further consideration. Evaluating next the tests on components, obvious significance is not achieved until the full first-order model, Model (11), is encountered, i.e., $L_{11-5}^2(6) = 819.77$, $p < .000$. Consequently, Model (5) and other members of the main-marginal family will not be considered at the expense of models in the first-order family because main-marginal models fit significantly more poorly than do first-order models. In sum, to this point, it appears that a model in the first-order family will approach acceptability for a symmetrical analysis.

---

[2] The BMDP/3F program was replaced by the 4F program in the 1981 version of the BMDP package. Aside from several new options in the 4F program, there are no major differences between these programs.

Selecting an Acceptable General Model

As an excuse to exercise our ability to select and interpret general models, let us assume for the moment that the inquiry of the McLean study was symmetrical. From knowledge gleaned from the screening table (Table 6.2), we would likely execute a computer run designed specifically to fit (a) first-order Models (6) through (11), (b) Models (5) and (12) so as to establish the outer boundaries of acceptability, and (c) the saturated model, a model that is always informative. The computer output will only provide residual chi-squares, but with these, component chi-squares can be hand calculated. We suggest that these statistics be organized in tabular form as shown in Table 6.3.

We approach the task of choosing a model without benefit of knowledge of the variables and substantive theory. Appreciating these limitations, let us begin by evaluating the residual L^2's given in Table 6.3, the results of which support rejection of Models (6), (7), and (8). Next, examination of component chi-squares confirms our earlier suspicion that second-order models, represented here by Model (12), also can be rejected as serious candidates for selection. Our search, therefore, will begin in earnest with Model (11) and will likely terminate by choosing a model that contains at least one first-order parameter. Before we begin, however, a few words should be said about the order in which first-order terms are entered in these models.

TABLE 6.3. Residual and Component Chi-squares for Selected Models Fitted to the McLean Data

Model No.	Residual			Component		
	df	L^2	*p*	*df*	L^2	*p*
(5)	11	829.83	.000			
(6)	10	714.24	.000	1	115.59	.000
(7)	9	628.32	.000	1	85.92	.000
(8)	8	610.01	.000	1	18.31	.000
(9)	7	18.96	.008	1	591.05	.000
(10)	6	12.47	.052	1	6.49	.025
(11)	5	10.06	.075	1	2.41	.121
(12)	4	8.80	.066	1	1.26	.262

Note. Models correspond by number to those presented in Table 6.1.

Recall from the last chapter that the order in which the three first-order terms appear in the full first-order model affects not only the magnitudes of respective components, but also the nature of the relationships (e.g., marginal or partial) associated with these components. Predictably, in this chapter, order of entry will also be found to influence the size and meaning of components but in a more complex manner. However, the magnitudes of specific effects (lambdas or taus) are *not* affected by order of entry. That is, for a model containing a given set of first-order terms, changing the order of appearance will not alter either the numerical value of effect parameters or the tests of significance (z tests) performed on these effects. In a sense then, statistical tests performed on component chi-squares are somewhat like the partial F tests performed on semipartial coefficients in multiple regression equations, where the latter are performed to determine whether a variable is making a significant contribution to the R^2. On the other hand, z tests performed on specific lambda effects in a log-linear model are analogous to t tests (or corresponding F tests) on specific coefficients (e.g., beta weights) in a multiple regression equation. It is hoped that a number of these points will be clarified during the course of the impending survey of the models presented in Table 6.3. The survey also has the purpose of further narrowing the range of terms that "should" appear in the most acceptable general model for the data gathered by McLean.

Model (11). Consider the component for this model: $L^2_{10-11}(1) = 2.41$, $p < .121$. In a linear model, this component is linked to λ^{cd}, the term that has been entered sixth or last in the array of first-order terms.[3] The magnitude of the component is related to the strength of a *full partial association*—the full partial association between Variables C and D in this case—and a test on this component will indicate whether or not the full partial association is statistically significant. To be more specific, the full partial here refers to the association between Variables C and D subsequent to effecting an adjustment for the potential confounding influences of Variable A, Variable B, and the association between Variables A and B which in the full first-order model, unlike other first-order models, is *not*

---

[3] Since an isomorphic relationship has been established between superscripts and subscripts for model terms, to promote economy of expression subscripts will be omitted whenever it is obvious from the context which term is being addressed.

independent of the association between C and D.[4] Thus, the statistical test of this component tends to be extremely conservative for it is a test of an association following removal of all factors that covary with it.

Notice in Table 6.3 that the test of the component for the *CD* association failed to result in significance, at least at a traditional level, despite the fact that respectable power has been afforded the test by the reasonably large sample. Even so, due to the conservative nature of the test and the fact that we are still engaged in a screening activity, λ^{cd} should not immediately be excluded from future consideration unless test results were decidedly nonsignificant, say $p > .25$. If we are to err during this tentative phase of model delimitation, it is best to err in the direction of inclusion rather than exclusion.

Additional insight can be obtained by examining the magnitudes of λ^{cd} effects in the full first-order model since these effects are invariant with respect to their ordering in this model. It turns out that the four effects are $\pm.093$, which when divided by their standard errors give z statistics greater than unity ($z = \pm1.29$). Though not impressive, effects with z statistics greater than unity are of interest, at least during this initial phase.

Before considering Model (10), examine Table 6.4, which contains: (a) component chi-squares that reflect the magnitude of marginal associations between variable combinations (the association between two variables subsequent to collapsing or summing over the two unspecified variables); (b) components that reflect full partial associations, obtained by entering a variable combination in the sixth position in a full first-order model and then contrasting that model with one that contains all terms belonging to the full first-order model except the term in question; and (c) z tests on estimated lambda effects contained in the full first-order model. The component for a marginal association is obtained by entering the corresponding term first or second in the array of first-order terms and then contrasting the five- (or six-) parameter model with an appropriate four- (or five-) parameter model. (Notice, for example, that the components for *AB* and *AC* given in Table 6.3 are the marginal

---

[4] Logically, it can be inferred that the association between A and B would be orthogonal to that between C and D. This inference is generally true; but not in the full first-order model in which multicollinearity abounds. Often, in fact, in the nonorthogonal environment of the full first-order model, the *AB* association can exert an influence on the *CD* association similar to that associated with suppressor variables in multiple regression.

TABLE 6.4. Tests of Marginal and Full Partial Associations and Tests of Lambda Parameters for the McLean Data

Effects	df	Marginal		Partial		
		L^2	p	L^2	p	z test*
AB	1	115.59	.000	30.05	.000	4.39
AC	1	85.92	.000	.83	.364	.77
AD	1	18.31	.000	9.37	.002	1.61
BC	1	675.81	.000	584.69	.000	15.42
BD	1	15.09	.000	.13	.723	.31
CD	1	16.93	.000	2.41	.121	1.29

*The absolute value of z statistics resulting from tests on lambda effects for the full first-order model, Model (11), produced by BMDP/3F, are contained in this column.

components listed for these combinations in Table 6.4.) And, as was just noted, the component reflecting a full partial association is obtained by entering the corresponding term in the final position in a full first-order model, creating an eleven-parameter model, then contrasting this model with a ten-parameter model where the latter is identical to the former except for the term in question. Upon request, the information presented in Table 6.4 is available on the output of the BMDP/3F and 4F computer programs. Needless to say, the availability of tests of marginal and full partials constitutes a salutary feature of these programs.

Model (10). The component $L^2_{9-10}(1) = 6.49$, $p < .025$ represents what we shall term a *half-partial association.* Here it is an association between Variables B and D subsequent to the partialling of influences exerted by Variable A, but not Variable C. Hence, when we refer to a half-partial association, control or adjustment has been exercised over only one variable (e.g., Variable A above), not both variables.

The nature of the association reflected by a component is dictated by preceding terms in a model. Consider specifically λ^{bd} and what might precede it in a model. If λ^{ac} is found prior to λ^{bd}, the association between A and C has no affect on the association between B and D, because they are independent of each other except in the highly dependent full first-order model. If, however, it should happen that *both* λ^{bc} and λ^{cd} have been entered prior to λ^{bd}, Variable C will be partialled from the association between B and D. Suppose, however, that λ^{ab} and λ^{ad} were entered before λ^{bd}, as is the case for current Model (10); what variable, if any, would be partialled out of the association between Variables B and D? Variable A is the answer because it is common to the prior associations involving Variables B and D. And since there must be at very least two common prior associations, documentation of half-partial associations can occur only for associations represented by terms entered third, fourth, or as we have just seen, fifth.

At this point, it appears that λ^{bd} is a likely candidate for inclusion in the final model. First appearances, however, can be deceiving, especially in log-linear work. A little additional probing, for example, will reveal two information items that cast serious suspicion on the importance of λ^{bd}. First, the component associated with its full partial association (see Table 6.4) will be found to be minuscule ($p < .72$). Second, the size of lambda effects in the full first-order model are also relatively unimpressive ($z = \pm.31$). Apparently, the association between Variables B and D diminishes markedly when Variable C is taken into account, a deduction that can be confirmed by testing the component for the half-partial association controlling for Variable C, the results of which are $L^2(1) = .91, p < .33$. Needless to say, even though λ^{bd} remains a candidate for inclusion, we should be most suspicious and critical of this term—a term that cannot stand independent of Variable C.

Model (9). Simply put, the component for this model is huge. What does it say about the association between B and C, however? The fact that λ^{ad} precedes λ^{bc} tells us nothing since the association between A and D is independent of that between B and C. But notice that two nonindependent terms involving Variable A—λ^{ab} and λ^{ac}—are present and prior to λ^{bc}. This arrangement points to a relationship between Variables B and C which is substantial even after the influences of Variable A have been taken into account.[5] Additional probing reveals that the component for the full partial association, $L^2(1) = 584.69$, is substantial and significant. Without doubt, λ^{bc} belongs in the model to be selected.

Model (8). The term in "third position" is similar to a term in the "fourth position" in that its component represents either a

---

[5] Of course, had it been found that λ^{bd} and λ^{cd} preceded λ^{bc}, a half-partial in which control was effected over Variable D would be indicated. Should the necessary conditions for the two types of half-partials not be present in a model, then the component for a term in the third, fourth, or fifth place would reflect either a marginal association or an association in which incomplete partialling has taken place; the latter type of association is nonsystematic and most difficult to interpret substantively.

half-partial, incomplete, or marginal association. Now, we know that if either λ^{ab} and λ^{bd}, or λ^{ac} and λ^{cd}, were to precede λ^{ad}, a half-partial would be indicated. Should λ^{ab} and λ^{cd} (or λ^{ac} and λ^{bd}) be found prior to λ^{ad}, the component would reflect an incomplete partial, and as indicated in footnote 5, incomplete partials are difficult to interpret substantively. But, the component in question reflects neither a half- nor an incomplete association; instead, by default, it reflects the strength of the marginal association between Variables A and D, which is statistically significant. In Table 6.4, significance also is seen for a full partial association, $L^2(1) = 9.37$, $p < .002$. And although short of significance, the tests on the λ^{ad}'s in the full first-order model produced respectable results ($z = \pm 1.61$). This term appears to be most promising.

Model (7). The component for a term in "second place" is readily interpreted: the component always reflects the strength of a marginal association. Here, the marginal association is between Variables A and C and this marginal association is significant: $L^2_{6-7}(1) = 85.92$, $p < .000$. The numerical value of this component would result from computing a likelihood-ratio chi-square on $[AC]$, subsequent to collapsing over Variables B and D. Table 6.4, however, informs us that the full partial association between A and C is not statistically significant ($p < .364$). Moreover, corresponding lambda effects in the full first-order model are quite small. In fact, the z statistics associated with these effects fail to exceed unity. All things considered, it is not likely that λ^{ac} will be found in the final model.

Model (6). The component linked to a first-order term that is entered first reflects the strength of a marginal association. This association is shown in Table 6.3 to be statistically significant. Also, both the complete partial association and z tests on the λ^{ab}'s are statistically significant. This term undoubtedly belongs in the final model.

What have we learned from our survey of first-order models? From all the evidence, it is clear that λ^{bc} and λ^{ab} are very important and belong in the model that will eventually be adopted. Though not as impressive, λ^{ad} is promising, and λ^{cd}, though questionable, should not be discarded yet. Candidates for exclusion appear to be λ^{bd} and λ^{ac}. Even though the former term showed a significant component in Table 6.3, suggesting an association between Variables B and D after adjustments are made for Variable A, when this association was viewed at respective levels of Variable C, it vanished. Furthermore, recall that the test on its component for a full partial fell far short of statistical significance ($p < .72$); and tests performed on lambda estimates were similarly unimpressive ($z = \pm.31$). The latter term, λ^{ac}, appears to be almost as impotent and for comparable reasons. (Parenthetically, investigation of half-partials revealed that when examined at respective levels of B, the relationship between A and C was nonsignificant, specifically $L^2(1) = 1.16$, $p < .20$.) Hence, at this point, we cautiously hypothesize that the most acceptable model will, in addition to main-marginal terms, contain λ^{bc}, λ^{ab}, λ^{ad}, and possibly λ^{cd}.

To assess and possibly fine-tune the hypothesized model, an additional computer run will be required. At very least, the following models should be fit:

$$\ln F_{ijkm} = \lambda + \lambda^a + \lambda^b + \lambda^c + \lambda^d + \lambda^{bc} + \lambda^{ab} + \lambda^{ad} \tag{6.2}$$

$$\ln F_{ijkm} = \lambda + \lambda^a + \lambda^b + \lambda^c + \lambda^d + \lambda^{bc} + \lambda^{ab} + \lambda^{ad} + \lambda^{cd} \tag{6.3}$$

Results, combined with those seen earlier for the full first-order model, are summarized in Table 6.5.

As expected, the full first-order model can be bypassed in favor of the nine-parameter model given by Eq. (6.3) since adoption of the latter more parsimonious model results in a minimal loss of ability to fit observed data, i.e., $L^2(2) = .90$, $p < .600$. The significant component aligned with the nine-parameter model, however, discourages us from dropping λ^{cd} and adopting Eq. (6.2). Thus, Eq. (6.3) is the most acceptable model, the best explanation, for the McLean data. Three strong relationships can be claimed, and a fourth—that between C and D—is present but may need to be advanced with qualification. Unfortunately, the substantive vacuum in which we have been operating makes it difficult to be more specific.

TABLE 6.5. Final Fit of General Models to McLean Data

Model	Marginals Fitted	df	Residual L^2	p	df	Component L^2	p
(6.2)	*BC, AB, AD.*	8	20.12	.010			
(6.3)	*BD, AB, AD, CD.*	7	10.96	.140	1	9.16	.002
(11)*	*AB, AC, AD, BC, BD, CD.*	5	10.06	.075	2	.90	.600

*This is the full first-order model described in Table 6.1 and subsequent tables.

Legitimate Logit Models

As the dimensionality of tables increase, so does the likelihood that inquiry is asymmetrical. In reality, the McLean study was asymmetrical. In this and other four-variable logit-model studies, the majority of general models shown in Table 6.1 do not apply. If we arbitrarily make Variable D the response variable, then the major requisite for a logit model is that it fits $[ABC]$. The model must contain λ_{ijk}^{abc} (or the corresponding tau parameter), for observed frequencies in the $[ABC]$ configuration can be the result of sampling or they can be deliberately fixed by the researcher. In either event, differing values of f_{ijk}^{abc}'s bear no relevance to patterns of proportional response over levels of the response variable, and consequently these differences must be controlled. Control can be exercised over irrelevant, potentially contaminating differences among f_{ijk}^{abc}'s by insisting that λ_{ijk}^{abc} be in any models where D is the logit variable. Main-marginal differences in $[D]$ also should not be permitted to influence proportional response to levels of D within other groupings; thus, it too should be a part of every logit model. Table 6.6. presents a hierarchical display of logit models for the layout under discussion.

Table 6.6 merits careful study. Due to their complexity, notice that we have abandoned the practice of numbering logit models according to the number of terms that they contain. The numbering system now reflects the number of interaction terms that involve the logit variable in the model. In addition, notice that every model in Table 6.6. fits the $[ABC]$ configuration and that only logit models (4), (5), (6), and (7) are in the listing of general models appearing in Table 6.1; the null logit model, Model (0), and Models (1), (2), and (3) have no counterpart in Table 6.1. Finally, note the bars over superscript d in all interaction terms. This is simply a convenient notational device popularized by Goodman to indicate that Variable D is the logit variable and, if superscripts do *not* carry a bar, to indicate corresponding effect designations as in the ANOVA.

In a number of respects, Table 6.6 is similar to the summary table for three-way ANOVA's, ANOVA's in which Variables A, B, and C are independent variables and Variable D is the dependent variable. Moreover, we will again conduct a logit-model analysis in a manner much like that of a three-way ANOVA. For our example, we will determine first whether there is a significant residuum observed about the F_{ijkm}'s provided by the null logit model. If the residuum about the null logit is significantly large, then the residual L^2 for this model will be partitioned into seven additive components, each component being associated with an effect that is potentially

TABLE 6.6. Hierarchical Logit Models for Four-Way Tables Where D Is the Logit Variable

Logit Model No.	Log-Linear Logit Models	Marginals Fitted
(0)*	$\lambda + a+b+c+d+ab+ac+bc+abc$	ABC, D.
(1)	$\lambda + a+b+c+d+ab+ac+bc+abc+ a\bar{d}$	ABC, AD.
(2)	$\lambda + a+b+c+d+ab+ac+bc+abc+ a\bar{d}+ b\bar{d}$	ABC, AD, BD.
(3)	$\lambda + a+b+c+d+ab+ac+bc+abc+ a\bar{d}+ b\bar{d}+ c\bar{d}$	ABC, AD, BD, CD.
(4)	$\lambda + a+b+c+d+ab+ac+bc+abc+ a\bar{d}+ b\bar{d}+ c\bar{d}+ ab\bar{d}$	ABC, ABD, CD.
(5)	$\lambda + a+b+c+d+ab+ac+bc+abc+ a\bar{d}+ b\bar{d}+ c\bar{d}+ ab\bar{d}+ ac\bar{d}$	ABC, ABD, ACD.
(6)	$\lambda + a+b+c+d+ab+ac+bc+abc+ a\bar{d}+ b\bar{d}+ c\bar{d}+ ab\bar{d}+ ac\bar{d}+ bc\bar{d}$	ABC, ABD, ACD, BCD.
(7)	$\lambda + a+b+c+d+ab+ac+bc+abc+ a\bar{d}+ b\bar{d}+ c\bar{d}+ ab\bar{d}+ ac\bar{d}+ bc\bar{d}+ abc\bar{d}$	ABCD.

Note. Constituent terms (parameter estimates) in models are denoted by their respective superscripts with the exception of the first term, λ.

*Numbers assigned to logit models in this table no longer reflect the number of constituent terms. They do, however, reflect the number of interaction terms that involve the logit variable.

responsible for the poor fit of the null logit model. Statistically significant components will indicate statistically significant main or interaction effects. Bear in mind, however, that the approach just described may also be viewed as selecting the most acceptable model as we did just a few pages previously. The model to be selected here, though, must be a legitimate logit model. Upon selection and fine tuning, an omnibus conclusion or conclusions based on the most acceptable model is advanced. Let us briefly survey the omnibus interpretations associated with the logit models that are seen in Table 6.6.

Interpretations of Logit Models

If it happens that Model (7), the saturated model, is chosen, second-order interaction effects, much like those seen in a three-way ANOVA, are indicated. That is, *simple* first-order interactions exist but they undergo modification upon examination at separate levels of the third variable. Granted, interpreting a second-order interaction may appear complicated. However, if Variable D, our assumed logit variable, is dichotomous, the interpretation is no more difficult than it is in a three-way ANOVA. Incidentally, adoption of the saturated model does not preclude the existence of overall main effects or overall first-order interaction effects.

Acceptance of Model (6), (5), or (4) indicates that at least one overall first-order interaction is present. For example, if Model (5) were deemed most acceptable, we would conclude that Variables A and C interact with respect to the logit variable. In addition, there does not appear to be an interaction between B and C or a second-order interaction because they are not in the model. But there is more. Because λ^{abd} precedes λ^{acd}, the interaction between Variables A and C is present over and beyond an interaction that might exist between Variables A and B. For the same reasons, if Model (6) were to be selected, we would conclude that there is a first-order interaction between Variables B and C subsequent to partialling out the interaction between A and B, and between A and C. Later in this chapter we will have the opportunity to select and follow-up a model in this family.

The family of main-effects models consists of Models (3), (2), and (1). Consider for a moment what it means to select Model (1). First, its adoption indicates that λ^{ad} is needed to achieve reasonable fit to observed data. Second, if our perspective were symmetrical, the presence of λ^{ad} would suggest a marginal association between

A and *D*. But our perspective is not symmetrical, and hence the presence of λ^{ad} suggests that the profile of proportional response over levels of Variable *D* is different when examined over levels of Variable *A*. Suppose Model (2) were to be chosen. From a symmetrical perspective, the presence of λ^{bd} would indicate a half-partial association, an association between *B* and *D* adjusting for Variable *A*. But, in a logit model, the need to incorporate λ^{bd} means that there are differences in the pattern of response to the logit variable when examined at different levels of Variable *B* even after adjustments have been made for main effects due to Variable *A*. Finally, suppose that Model (3) happend to be the model of choice. In the symmetrical, λ^{cd} would point to a full partial association between *C* and *D*; but in the asymmetrical, this can be translated to mean that main effects are present among levels of Variable *C* subsequent to correcting for the main effects of both Variables *A* and *B*. Again, order of variable entry is important to interpretation and merits careful a priori consideration.

The remaining model in Table 6.6 is the null logit model, appropriately designated Model (0). If this model fits observed data reasonably well, we must conclude that there are no effects of sufficient magnitude to be entertained. Adoption of the null model is analogous to performing a three-way ANOVA and discovering that, of the seven omnibus *F* tests, not one has achieved statistical significance.

Let us make one more attempt to relate a logit-model analysis to an ANOVA. In the ANOVA the sum of squared deviations about the grand mean (i.e., total sum of squares) is calculated and then partitioned into specific components (e.g., SS_A, SS_B, etc.) associated with factors (e.g., Variable *A*, *B*, etc.) that explain variability about the grand mean. In a logit-model analysis, we attempt to explain the discrepancy between observed cell frequencies and those offered by the null logit model. The size of the discrepancy is reflected by the size of the residual chi-square for Model (0). This residual chi-square can then be partitioned into specific components (e.g., L_{0-1}^2, L_{1-2}^2, etc.) associated with factors (e.g., Variable *A*, *B*, etc.) that appear to be promoting discrepancy about the expected cell frequencies generated by the null. Moreover, in the ANOVA, if cell *n*'s are unequal and disproportionate, component sums of squares are not independent; they are nonorthogonal, and ANOVA methods based on comparisons between regression models need to be evoked. Methods not dissimilar to those used for unequal *n* ANOVAs will be illustrated in the next section.

AN EXAMPLE OF A LOGIT–MODEL ANALYSIS

McLean, cited previously, sought to investigate the graduation rates of black and white students in two southern universities: historically, one was black, the other white. Research indicated that, in general, black students tend to complete their undergraduate degree programs at a lower rate than white students. McLean, however, suspected that the differential rate of program completion was moderated in part by the type of institution attended. Specifically, he suspected that differences between black and white completion would be more pronounced in historically white universities than in historically black universities, where in the latter differences may not exist.

Selected for study were 896 first-time, full-time students who enrolled in degree programs in either the white or the black university during the Fall of 1975. Due to the small numbers, during the Autumn all blacks entering the white university were included in the sample. Remaining black and white sample members at respective universities were drawn at random. Data pertaining to high school performance were used in regression equations to predict college success wherefore samples within universities were bifurcated into a group that had prediction scores above the average (termed high-ability students by McLean) and a group with below-average predictors (low-ability students). At the end of the fourth academic year (Spring 1979), it was determined whether each sample member had graduated or had failed to graduate during the four-year period. Thus, this research involved the study of three dichotomous explanatory variables and a dichotomous response variable, completion status. The variables were:

Variable A: Ability Groups
A_1 = high ($f_1^a = 447$)
A_2 = low ($f_2^a = 449$)

Variable B: Type of University
B_1 = historically white ($f_1^b = 426$)
B_2 = historically black ($f_2^b = 470$)

Variable C: Race of Student
C_1 = black ($f_1^c = 492$)
C_2 = white ($f_2^c = 404$)

Variable D: Completion Status, Spring 1979
D_1 = graduated ($f_1^d = 308$)
D_2 = did not graduate ($f_2^d = 588$)

TABLE 6.7. The McLean Data: Observed Frequencies by Race, Type of University, Ability Level, and Completion Status

Race	University	Ability	Completion Status	
			Graduation	Nongraduation
Black	White	High	10	22
		Low	4	18
	Black	High	55	90
		Low	71	222
White	White	High	114	146
		Low	46	66
	Black	High	5	5
		Low	3	19

The elementary cell frequencies reported by McLean are reproduced in Table 6.7. Prior to considering a logit-model analysis of these data, it is of interest to note the results of analyses of racial group by completion status when these analyses are performed separately within institutions. If the tabulations of the 426 students matriculating at the historically white institution are organized into a fourfold table defined by Race (black and white) and Completion Status (graduation and nongraduation), and a Pearsonian chi-square is calculated, significance is observed; specifically, $\chi^2(1) = 5.76$, $p < .05$. At the white university, fewer black students than expected under the null graduated during the period under study. In contrast, a similar analysis performed on the 470 observations peculiar to the historically black university failed to attain significance, $\chi^2(1) = .64$. Thus a fragmentary, insular analysis of these data would tend to support McLean's initial suspicion that Race and Type of University *interact* with respect to completion status.

A summary of the fit of models, arranged in the hierarchy shown in Table 6.6, is presented on the left side of Table 6.8.[6] Note first that the null model does not fit observed data well: $L_0^2(7) = 36.00$, $p < .00$. Therefore, we have a "green light" to

---

[6] The results of Table 6.8 are not identical to those reported by McLean (1980, pp. 71-72). McLean used BMDP/3F but in doing so specified "delta = .5," used when tables contain sampling zeros, despite the fact that sampling zeros were not present in his data. The results in Table 6.8 are those analyzed by this writer without the delta option.

TABLE 6.8. Summary of Logit-Model Analysis on McLean Data

Model*	Hierarchical Components Source	L^2	df	p	Full-Partial Components Source	L^2	df	p
(0)	Null-Logit	36.00	7	.000				
(1)	Due to A	18.31	1	.007	Due to A : $B \& C$	9.37	1	.002
(2)	Due to B : $A^\dagger$	6.49	1	.011	Due to B : $A \& C$	.13	1	.723
(3)	Due to C : $A \& B$	2.40	1	.122	Due to C : $B \& C$	2.40	1	.122
(4)	Due to AB	3.06	1	.108	Due to AB : $AC \& BC$	1.26	1	.262
(5)	Due to AC : AB	.32	1	.540	Due to AC : $AB \& BC$	.02	1	.886
(6)	Due to BC : $AB \& AC$	2.89	1	.089	Due to BC : $AB \& AC$	2.89	1	.089
(7)	Due to ABC	2.53	1	.112				

*The hierarchical models specified in this table are those defined in Table 6.6.
†Read: Variation from the null-logit model due to effects of Variable B, subsequent to affecting adjustments for Variable A.

proceed to specific sources to see which effects are most responsible
for the residuum observed about the null. Starting at the bottom,
as is also common when reading an ANOVA table, in the left-hand
portion of Table 6.8 we find that the component for second-order
interaction and the three first-order components are not significant
at the .05 level. This means that interaction between or among
explanatory variables can be ruled out. Somewhat of a surprise was
the fact that the interaction between B (University) and C (Race)
was not significant, i.e., $L^2_{5-6}(1) = 2.89$, $p < .09$. After all, McLean
suspected that retention by race was a function of the type of
institution and, furthermore, the Pearsonian chi-squares mentioned
a moment ago for the white university ($\chi^2 = 5.76$) and the black
university ($\chi^2 = .64$) suggested that the retention rates of blacks
and whites were different at respective institutions. But the more
comprehensive analysis has shown that the interaction between
B and C is not prominent if the main effects of ability and other
first-order interaction effects are taken into account. Incidentally,
had McLean chosen not to effect the adjustments just mentioned,
but instead had entered λ^{bcd} in Model (4) in place of λ^{abd}, the
resultant component (a component not adjusted for the remaining
two first-order interactions) would still fall short of statistical significance, i.e., $L^2(1) = 3.08$, $p < .08$.

Examining main effects in Table 6.8 reveals that significance
can be claimed for ability level ($p < .007$) and universities subsequent
to correcting for levels of ability ($p < .011$). However, if components reflecting full partials are examined—components reflecting
full partials are given to the right in Table 6.8—university effects
become minuscule. In the final analysis, McLean could only support
the common-sense finding that high-ability students are more likely
to graduate than low-ability students. But fortunately, because of
his choice of method, McLean minimized the chances of positing
a finding that is spurious when other variables are taken into account,
an advantage of log-linear analysis that should not be minimized.

EXPLICATING INTERACTION IN
A LOGIT–MODEL ANALYSIS

Recall from the example that McLean believed that he would be able
to document an interaction of given form between universities
(Variable B) and student's race (Variable C). To do this, λ^{bcd} would

first have to be statistically prominent in the logit model deemed most acceptable. Subsequently, the pattern of first-order interaction between Variables B and C with respect to logit Variable D—an interaction that we shall term a *first-order-logit interaction*—would need to be examined to see if it were consistent in form with that which had been hypothesized. But neither of these events occurred since the component chi-square for the term in question was not observed to be statistically significant at the .05 level: $L^2(1) = 2.89$, $p < .089$. The interaction, however, did "approach" significance, and since we are both moderately familiar with the McLean study and keen to learn how to handle a first-order-logit interaction, for the sake of pedagogy let us assume that the interaction of interest turned out to be significant so that it can be pursued.

Examining the Size and Pattern of Effect Parameters

Assuming $\lambda^{b\bar{c}\bar{d}}$ to be significant, we know only that black and white students have different rates of graduation at the two different universities. Obviously the nature of the "differences between differences" needs to be studied. Some of the same steps that were taken to explicate the main effects discovered in the reflective teaching example of the last chapter will be taken here to follow-up the first-order-logit interaction. Recall that these steps were: (a) identification of the most appropriate logit model to serve as the follow-up model and (b) examination of the size, direction, and pattern of relevant effects ($\lambda^{b c d}$'s in the present instance) in the follow-up model and, for good measure, in the saturated model.

In terms of selecting a follow-up model, a strong case can be made for the adoption of a 13-parameter model that fits configurations $[ABC]$, $[BCD]$, and $[AD]$. For reasons previously discussed, the fitting of $[ABC]$ is mandatory for all logit models pertaining to this example. In addition, $[BCD]$ is needed because it represents the first-order-logit interaction that we are assuming to be significant, and $[AD]$ should be fitted because of the prominent and significant main-effects for Variable A seen in the McLean study. In an effort to lend additional support to the selection of the 13-parameter model, an analysis was undertaken of the residuum about the F_{ijkm}'s produced by this model along lines discussed in Chapter 4. There were no instances where the standardized or Freeman-Tukey deviates either achieved or suspiciously approached statistical significance (at the .05 level), nor were there apparent patterns in the algebraic

TABLE 6.9. Estimates of Logit Interaction Effects between Variables B and C in McLean's Data

Var. B, Univer.	Var. C, Race	Accepted Model		Saturated Model	
		Grad.	Nongrad.	Grad.	Nongrad.
White	Black	−.114*	.114	−.119†	.119
	White	.114	−.114	.119	−.119
Black	Black	.114	−.114	.119	−.119
	White	−.114	.114	−.119	.119

*The z tests on effects associated with the most acceptable logit model were $z = \pm1.58$, $p < .12$, two tail.

†z tests on effects from the saturated model were $z = \pm1.64$, $p < .10$, two tail.

signs of the deviates that might indicate the absence of an important term in the model.[7]

Having selected a follow-up model, estimates of effect parameters need to be requested as computer output. The $\lambda^{bc\bar{d}}$'s associated with the logit model deemed most appropriate for McLean's data and those associated with the saturated model are displayed in Table 6.9.

Before the $\lambda^{bc\bar{d}}$'s in Table 6.9. are examined, let us be sure that we understand what they mean. $\lambda^{bc\bar{d}}$, an unspecified first-order-logit interaction (or a second-order interaction in a general model) may be defined verbally as a difference between the log of the geometric mean of F_{ijkm}'s in the jkmth combination of $[BCD]$ and the log of the grand geometric mean of F_{ijkm}'s (i.e., $\ln \bar{G} = \lambda$), subsequent to correcting that difference six times: three times for main-marginal effects and three times for relevant two-variable associations or interactions. Defined mathematically,

$$\lambda^{bc\bar{d}}_{jkm} = (\ln \bar{G}^{bc\bar{d}}_{jkm} - \ln \bar{G})$$
$$-\lambda^{b} - \lambda^{c} - \lambda^{d} - \lambda^{bc} - \lambda^{b\bar{d}} - \lambda^{c\bar{d}} \tag{6.4}$$

Because different models generate different F_{ijkm}'s, the numerical

---

[7]Bishop et al. (1975, pp. 137–139) illustrate that patterns in the algebraic signs of deviates can provide insights into why a particular model is not fitting data well, and what new term might be added to improve the fit.

values of ln $\bar{G}^{bcd}$ and ln $\bar{G}$ can be expected to vary, and hence the numerical values of λ^{bcd}'s will often vary from model to model. But more important is the realization that for a given model, and from a logit-model perspective, the λ^{bcd}'s in Table 6.9 constitute relatively pure estimates of the interaction between Variable B (University) and Variable C (Race), estimates unaffected by the logit main effects of Variables B and C. An examination of λ^{bcd}'s, therefore, will provide us with an intemerate view of first-order-logit interaction. Such an examination becomes the first serious step in our endeavors to explicate the interaction.

The λ^{bcd}'s belonging to the follow-up model and the saturated model, plus the z statistics resulting from tests on these estimates, are seen in Table 6.9 to be very similar. Since there is only one *basic* λ^{bcd} effect, in effect only one independent test is being run for each model. That aside, the results of all testing failed to achieve significance at the .05 level, an outcome to be expected considering that the composite test reported earlier in Table 6.8 did not satisfy the .05 criterion, i.e., $L^2(1) = 2.89$, $p < .089$. But since we are still assuming that the interactions are worthy of interpretation, we note that the pattern exhibited by λ^{bcd}'s permits us to say that at the historically white university, a greater proportion of black students tend not to graduate "on time." Or, put differently, at the historically black school, a higher proportion of white students in attendance tend not to graduate on time. Strictly speaking, however, these conclusions are only appropriate subsequent to making adjustments for logit main effects—subsequent to making an adjustment for the fact that a higher percentage of students (black and white) in the black university did not graduate on time when compared to all students attending the predominantly white university (71% vs. 59%), and the fact that a higher percentage of black students, as compared to whites, combined over both schools did not graduate on time (72% vs. 58%).

Examining Binomial Logits

While not meaning to depreciate the value of our work to this point, remember that in the actual conduct of inquiry we are apt to be more interested in the aggregate impact of effects on groups rather than in the analysis of specific effects per se. For example, in the McLean study, subsequent to analyzing pure interactions we would likely want to examine the combined influence of interaction and main effects as it affects the two student groups in the two universities. Combined influence may be examined and communicated to others through the use of either (a) a graph of *binomial logits* which has the appearance of a cell-mean graph frequently used in the ANOVA to depict interaction, or (b) a graph of the proportional response to levels of the logit variable within relevant groups. To effect a comprehensive assessment of interaction, the use of at least one of these graphical procedures is recommended. Constructing a graph of binomial logits will be illustrated here; a discussion of the second graphical procedure will be postponed until later in this chapter where we will again face the task of explicating a first-order-logit interaction, but the task will be of even greater challenge because the response variable will be a polytomy.

The graphing of binomial logits is most advisable when the response variable is a dichotomy. The basic idea is to reduce a three-variable configuration to a two-variable configuration and to transform the qualitative dichotomous response to a single variable measured on a metric scale. In terms of the McLean study, the idea is to reduce $[BCD]$ to $[BC]$ and, for each of the cells in $[BC]$, to transform Variable D to a metric measure. The measure, called the binomial logit (or half-logit) is symbolized by ψ and it can be defined by

$$\psi_{1/2}^{b c \bar{d}} = \ln \left(f_{jk1}^{b c \bar{d}} / f_{jk2}^{b c \bar{d}} \right)^{1/2} \tag{6.5}$$

which is not as intimidating a definition as it may appear to be at first glance. As we shall see, Eq. (6.5) yields unbounded metric values that can be plotted like cell means in the ANOVA.

We again use data gathered by McLean, and shown in Table 6.10, to illustrate. The reader should undertake to find in the table

TABLE 6.10. Odds and Binomial Logits from Response Variable D as Observed in Configuration (BC) of McLean's Data

University	Race	D_1	D_2	$\omega_{1/2}$	$\ln \omega_{1/2}$	$\psi_{1/2}$
White	Black	14	40	.350	−1.049	−.525
	White	160	212	.755	− .281	−.141
Black	Black	126	312	.404	− .907	−.453
	White	8	24	.333	−1.099	−.550

Note. Observed odds, logits, and binomial logits reflect the odds of being classified in D_1 (on-time graduation) as opposed to D_2 (did not graduate).

the $f^{b c \bar{d}}$'s observed in [BCD]. Note that the column headed $\omega_{1/2}$ contains *conditional odds*, the observed odds of being classified in D_1 (graduated) as opposed to D_2 (did not graduate) for each combination in [BC]. If our point of reference were the odds of being in D_2 as opposed to D_1, the column heading would read $\omega_{2/1}$. Notice how the dichotomous response variable has been converted into a univariate metric variable for reduced configuration [BC].

Incidentally, Goodman and many other prominent developers of log-linear analysis have a penchant for viewing qualitative response in terms of odds and odds ratios, for they believe that odds constitute the least ambiguous vehicles for understanding and interpreting log-linear results (cf. Goodman, 1978). Proponents of the use of odds to express response also make extensive use of log-linear models that yield *expected odds*—as opposed to the models used extensively herein that yield expected cell frequencies. In all candor, there is much to recommend odds as a vehicle for comprehension and interpretation. They have received little attention in this book, however, because a major objective of this work is to expose more fully, even to the point of exploitation, the many similarities between the familiar system of analysis known as ANOVA and the emerging body of method known as log-linear analysis. Therefore, the approach followed in these chapters, while not deliberately intending to eschew the use of odds, in effect does so due to the deliberate emphasis given to perceiving qualitative response in terms of ANOVA-like effects.

We did, however, briefly discuss conditional odds in Chapter 4, where we learned that they possess a minor, yet vexatious, problem: their distribution is not symmetric about a mid-point of unity. That is, the distribution of ω's is nonsymmetric about $\omega = 1.00$. We overcome this problem, as before, by working instead with the natural logs of the odds. The logs of the $\omega_{1/2}$'s can be found in Table 6.10.

Finally, in the right-most column of Table 6.10 are located the binomial logits. Frequently, Greek psi is used as a general symbol for these special logits, logits which are obtained by simply dividing the logs of the odds by 2. They can also be computed from Eq. (6.5), a fact that should be verified and an equation that now should have increased meaning. The $\psi_{1/2}$'s are negative in this table because the odds of on-time graduation are less than unity for the four groupings of students. On the other hand, by reversing the perspective and using Eq. (6.5) to calculate $\psi_{2/1}$'s instead of $\psi_{1/2}$'s, we would find no change in the numerical values of the binomial logits but their algebraic signs would all be reversed: they would all be positive.

Another interesting characteristic of binomial logits is that they are equivalent to corresponding lambda effects in the absence of confounding. For the McLean situation, if it had turned out that there were no differences in $[D]$ (i.e., $f_1^d = f_2^d$), and no logit main effects for either Variable B or Variable C, then the binomial logits would be equal to the $\lambda^{bc\bar{d}}$'s. To the extent to which these effects are operative, however, binomial logits and lambdas will not be equivalent. This is because the former will reflect the confounding of the aforementioned effects and interaction, while the latter will reflect pure interaction. Nevertheless, the confounded outcome often is of greater practical interest, and hence the plotting of binomial logits has been offered as a meaningful device to use subsequent to detecting first-order-logit interaction (Marks, 1975).

Figure 6.1 contains a graphical display of binomial logits by Race of Students as a function of Type of University. The pattern of interaction is disordinal but it is not symmetric, in the sense that the difference between black and white students is greater at the

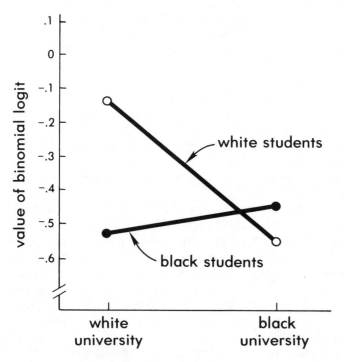

Figure 6.1. Plot of binomial logits by Race of Students and Type of University.

white university than it is at the black university. At the white university, for example, proportionately more white students are seen to have graduated on time. That is, the odds of graduating on time were higher for white than for black students. Though not as pronounced, at the black university, the opposite is seen. But, aside from a degree of distortion or lack of symmetry in the pattern of interaction, substantive conclusions are similar to those advanced following an examination of pure interaction effects in Table 6.9.

The binomial logits appearing in Figure 6.1 are susceptible to the confounding influences noted earlier. Notice, for example, where the pattern or configuration is located on the graph. Notice particularly where the configuration is centered: it is entirely in the negative realm, centered well below zero, centered well below the point that is indicative of even odds. Differences in the main-marginals of Variable D (i.e., $f_1^d = 308$, while $f_2^d = 588$) are responsible for this. If instead of plotting $\psi_{1/2}^{bcd}$'s, we instead decided to plot the $\psi_{2/1}^{bcd}$'s, the configuration would reside in the positive realm.

Distortion in the symmetry of the configuration is due to the confounding of logit main effects for Variables B and C. Observe that the interaction configuration appears to be "tilted" in a downward direction when viewed from left to right. In part this tilt reflects the fact that proportionately more students in the historically white institution graduated on time. Also, the fact that a greater proportion of white students (as opposed to black students) graduated on time is causing the difference between blacks and whites to be greater at the white university than it is at the historically black institution.

In sum, there are a number of factors mixed in with interaction effects that determine the algebraic values of binomial logits. When graphed, therefore, binomial logits reveal net influences and reflect a substantively meaningful picture of results. Incidentally, the pattern of results depicted in Figure 6.1 comes close to the initial suspicions held by McLean. Despite respectable sample size, however, he was unable to support statistically the pattern of results that we have just discussed. Nevertheless, in practice there is much to recommend the plotting of binomial logits. Graphs depicting proportional response within groups to levels of the logit variable, an example of which we shall see in the next section, are also to be recommended.

FIVE-DIMENSIONAL TABLES

The preceding principles are generalizable to tables of high dimensionality. An illustration of a five-variable problem in which we see a polytomous response variable is offered in this section. The illustration is based in part on the work of Peters (1981) although both the design and resultant data have been altered somewhat to effect a more complex result.

A Study of Preferred Styles of Inquiry

Let us say that the sample for this working example consisted of 296 graduate students pursuing doctoral degrees in the social sciences at a large midwestern university. Participating students were administered the Myers-Briggs Type Indicator (MBTI) (Myers, 1962), a well-established clinical and research instrument that can be used to classify examinees on four personality dimensions, the majority of which were advanced by the Swiss psychiatrist Carl Gustav Jung. The four dichotomous classifications constitute four explanatory variables in the emended Peters' study. Specifically, following administration of the MBTI, subjects were cross-classified on the basis of the following:

1. Variable A: Extraversion (E) or Introversion (I)
   A_1 = classified as an E ($f_1^a = 134$)
   A_2 = classified as an I ($f_2^a = 162$)

2. Variable B: Intuition (N) or Sensing (S)
   B_1 = classified as an N ($f_1^b = 123$)
   B_2 = classified as an S ($f_2^b = 173$)

3. Variable C: Thinking (T) or Feeling (F)
   C_1 = classified as a T ($f_1^c = 150$)
   C_2 = classified as an F ($f_2^c = 146$)

4. Variable D: Judging (J) or Perceiving (P)
   D_1 = classified as a J ($f_1^d = 139$)
   D_2 = classified as a P ($f_2^d = 157$)

Peters, attempting to subject to test a hypothesis advanced by Mitroff and Kilmann (1978), sought to determine whether subjects' personality type, as measured on the MBTI, influenced their preference for certain styles of knowledge production. To assess subject preference, four published articles that address the topic of faculty development were identified. Each article approached the topic from a distinct "style of inquiry." Subjects were asked to select one of the four articles, namely the article that in their view possessed the "greatest potential for contributing to educational understanding."

Choice of article, therefore, was the response variable. A brief description of this polytomous variable follows:

5. Variable E: Choice of preferred style of inquiry

E_1 = choice of the ST article, an article that represented the ST mode of inquiry as described in the work of Mitroff and Kilmann. This article approached the topic of faculty development in inductive, molecular, and experimental manner and was redolent with tabular material and quantitative analyses.

E_2 = choice of the NF article, an article that assumed an "action-oriented" approach to the common topic through the implementation, but not development, of a theoretical model.

E_3 = choice of the NT article, which approached the topic from a theoretical perspective. It represented an effort to advance a conceptual framework for the understanding of faculty growth.

E_4 = choice of the SF article, an article that approached the topic in a highly personal, concrete, affective manner with little recourse to theory or research.

It is, of course, well known that if the response variable for the foregoing study were measured on either an interval or a ratio scale, the appropriate analysis would be a four-way ANOVA. But Variable E is neither interval nor ratio; it is decidedly qualitative. Therefore, a four-variable logit-model analysis, an analysis that parallels a four-way ANOVA, is appropriate. Further, as a point of departure, realize that all logit models that apply to this study must at least fit [$ABCD$] and [E]. These requisite fittings will make adjustments for potential inequalities in f^{abcd}'s and f^e's, inequalities that are not only irrelevant to logit-variable response but if left uncontrolled would distort logit-variable response. Hence, notice that all logit models in the hierarchical arrangement presented in Table 6.11 contain these two requisite fittings.

One last attempt at an intuitive explanation of logit models seems warranted. Consider first Model (0), the null logit model, a model that contains the two requisite fittings mentioned above and thus, in its extended form, contains 17 effect parameters none of which stand for an interaction involving Variable E. Using maximum-likelihood procedures, expected frequencies for the 64 elementary

TABLE 6.11. A Hierarchical Arrangement of Logit Models for Four Explanatory Variables

Model No.	Source of Variance about the Null Model		Marginal Configuration Fitted
(0)	Null Logit		$ABCD, E.$
(1)	A	$: A$	$ABCD, AE.$
(2)	B	$: A \& B$	$ABCD, AE, BE.$
(3)	C	$: A, B \& C$	$ABCD, AE, BE, CE.$
(4)	D	$: A, B \& C$	$ABCD, AE, BE, CE, DE.$
(5)	AB	$: ME*$	$ABCD, ABE, CE, DE.$
(6)	AC	$: ME \& AB$	$ABCD, ABE, ACE, DE.$
(7)	AD	$: ME, AB \& AC$	$ABCD, ABE, ACE, ADE.$
(8)	BC	$: ME, AB, AC \& AD$	$ABCD, ABE, ACE, ADE, BCE.$
(9)	BD	$: ME, AB, AC, AD \& BC$	$ABCD, ABE, ACE, ADE, BCE, BDE.$
(10)	CD	$: ME, AB, AC, AD, BC \& BD$	$ABCD, ABE, ACE, ADE, BCE, BDE, CDE.$
(11)	ABC	$: ME \& FOE^{\dagger}$	$ABCD, ABCE, ADE, BDE, CDE.$
(12)	ABD	$: ME, FOE \& ABC$	$ABCD, ABCE, ABDE, CDE.$
(13)	ACD	$: ME, FOE, ABC \& ABD$	$ABCD, ABCE, ABDE, ACDE.$
(14)	BDC	$: ME, FOE, ABC, ABD \& ACD$	$ABCD, ABCE, ABDE, ACDE, BCDE.$
(15)	$ABCD$		$ABCDE.$

Note. Variable E is assumed to be the response variable.
*The letters ME stand for all main effects as specified in Models (1) through (4).
†The letters FOE stand for all first-order interaction effects as specified in Models (5) through (10).

cells of the $2 \times 2 \times 2 \times 2 \times 4$ table are generated. The F^{abcde}'s given by the null logit model represent a state of affairs that assumes that frequencies in all configurations other than $[ABCD]$ and $[E]$ are equiprobable. Therefore, should Model (0) be deemed to fit observed data well, differences between respective estimated and observed elementary cell frequencies would be assumed to be due to chance, an assumption that precludes the mention of significant effects between or among explanatory variables.

Suppose, however, that, in addition to $[ABCD]$, it was found that $[CE]$ was needed to achieve reasonable congruence between expected and actual frequencies. (Incidentally, this particular model is not shown in the table of logit models.) What can this be taken to mean? Given the current problem, it would mean that if we focused our attention on any one of the 16 variable combinations in $[ABCD]$, and if within this selected combination we examined patterns of proportional response over levels of Variable E, a statistically meaningful difference in response patterns would be seen between subjects classified as T's and as F's on Variable C. Moreover, since we are assuming that no additional parameters are required in the model, the difference between T's and F's relative to their patterns of response to the logit variable is similar, at least within chance expectation, within all remaining 15 combinations in $[ABCD]$. In short, adoption of this model would indicate significant main effects for Variable C.

Let us extend this line of thinking by momentarily assuming that, in addition to $[ABCD]$, $[BCE]$ was needed to achieve reasonable fit. Fitting $[BCE]$, of course, also implies the fitting of $[BE]$ and $[CE]$ irrespective of whether these configurations contain beyond-chance differences. In any event, the need to incorporate $[BCE]$ in the model is an acknowledgment that there are discernable differences among the f^{bce}'s. Therefore, if a particular variable combination in $[ABCD]$ is selected again for study, a difference between T's and F's with respect to their pattern of proportional response to the logit variable will be observed. But, because of the indicated interaction between Variables B and C, unlike before we will not expect the difference between patterns to be similar in the 15 remaining combinations of $[ABCD]$. To be more specific, the *simple* main effects between T's and F's viewed at the first level of Variable B will be different from the simple main effects viewed at the second level of Variable B. Looking ahead, it is precisely this type of interaction that will be encountered in the Peters' example.

Returning to the study, data that were generated by this writer were analyzed by the 3F program in the BMDP series (1979), the

TABLE 6.12. Summary of Logit-Model Analysis of MBTI Data

Model No.	Residual			Component		
	df	L^2	*p*	*df*	L^2	*p*
(0)	45	113.75	.00			
(1)	42	111.45	.00	3	2.30	.51
(2)	39	110.60	.00	3	.85	.84
(3)	36	27.58	.84	3	82.02	.00
(4)	33	27.47	.74	3	.11	.99
(5)	30	25.23	.71	3	2.24	.52
(6)	27	21.22	.78	3	4.01	.27
(7)	24	20.93	.64	3	.29	.96
(8)	21	11.52	.95	3	9.41	.02
(9)	18	5.75	.99	3	5.77	.12
(10)	15	5.47	.99	3	.28	.96
(11)	12	3.75	.99	3	1.72	.64
(12)	9	2.88	.97	3	.87	.83
(13)	6	2.25	.90	3	.63	.89
(14)	3	1.53	.68	3	.72	.87
(15)	0	0.00	—	3	1.53	.68

Note. Models correspond by number to those appearing in Table 6.11.

precursor to BMDP/4F (1981). Table 6.12 contains a summary of results, where we look first to see how well the null logit model fits observed data. The residual chi-square for this model was substantial: $L^2(45) = 113.75$, $p < .00$. Clearly, the poorness-of-fit offered by this model invites an examination of component chi-squares for the purpose of identifying effects most responsible for the significant residual discrepancy between expected frequencies of the null and observed frequencies.

Further examination of Table 6.12 reveals that only two component chi-squares are statistically significant at the .05 level: the component associated with main effects for Variable C (subsequent to adjusting for Variables A and B) and the component for the interaction between Variables B and C (after adjustments for all main effects and three first-order interactions).

But before we follow-up the two sets of effects that are obviously significant, at this tentative stage we should not arbitrarily dismiss the interaction between Variables B and D for it is not altogether inconsiderable ($p < .12$). Thus, prior to the final selection of a follow-up model, the 22-parameter model defined by the fitting of [ABCD], [BCE], and [BDE] was compared to a 20-parameter model defined by fitting [ABCD] and [BCD]. The residual chi-square for the 22-parameter model was $L^2(30) = 12.56$, $p < .99$; for the more restricted 20-parameter model it was $L^2(36) = 18.83$, $p < .99$. Contrasting respective residual chi-squares gave a component of 6.27 which when taken to a table of chi-squares distributed on six degrees of freedom was not significant ($p > .30$). An examination of standardized residuals and Freeman-Tukey deviates also failed to provide evidence that would suggest that the more restricted model was not an acceptable fit. The 20-parameter model, therefore, was chosen over the 22-parameter model for further study of the logit main effects of Variable C (T's versus F's) and the first-order-logit interaction between Variables B and C.

Following-up Main Effects

Having established that subjects who were classified as either T's or F's differed in their preference for at least one article, much can be learned about the specific nature of this difference, or differences, by examining relevant $\lambda^{c\bar{e}}$'s. Hence, in Table 6.13, we can find among other statistics the $\lambda^{c\bar{e}}$'s associated with the 20-parameter follow-up model. (Corresponding lambdas associated with the saturated model were found to be sufficiently similar to those given in the table and thus their presentation was deemed unnecessary.) The algebraic signs of the lambda effects in question suggest that proportionately more T's (as compared to F's) indicated a preference for the ST article and the NT article. In contrast, in greater proportions, F's preferred the NF and SF articles. Additional support for

TABLE 6.13. Estimates of Main Effects for the Dimensions of Thinking (T's) versus Feeling (F's)

Article Preference	[E]	Thinking			Feeling			
		λ	f^{ce}	p	λ	f^{ce}	p	z test
ST	61	.465	48	.32	−.465	13	.09	3.87
NF	71	−.675	14	.09	.675	57	.39	5.80
NT	70	.610	58	.39	−.610	12	.08	5.09
SF	94	−.400	30	.20	.400	64	.44	3.88
Totals	296	0.000	150	1.00	0.000	146	1.00	

these observations can be found in the table. Specifically, both the f^{ce}'s and patterns of proportional response to articles computed *within* groups of T's and F's are compatible with the stated observation. Technically speaking, however, we know that the $\lambda^{c\bar{e}}$'s and the profiles of within-group proportional response do not convey exactly the same message. The former are *effects* (literally, difference between means) that have been adjusted for a very slight main-marginal inequality in $[C]$ and again for differences in $[E]$ that are shown in the second column of the table. The latter, the profiles of proportional response, are only adjusted for main-marginal differences in $[C]$—an adjustment that is a consequent of computing proportions such that they sum to unity within groups. The within-group proportions are influenced by main-marginal differences in $[E]$, however. Nevertheless, both views of these logit main effects are meaningful. The lambda effects depict a pure, unconfounded picture of simple main effects for T's and F's with respect to their article preference, while within-group proportions reflect simultaneously both simple main effects and overall article preference.

So far, our discussion of differences between T's and F's has been confined to a descriptive level. Simple main effects have yet to be subjected to statistical test, namely to z tests. These tests, which conform to conventional unit normal theory, are provided by most log-linear computer programs, and the results for our study are shown in Table 6.13. To claim statistical significance at say the .05 level, two tail, we know that conventional practice holds that the computed z statistic must exceed the 97.5-centile value (critical value) of 1.96. Notice that each test reported in Table 6.13 satisfies this criterion.

We can no longer ignore the problem of repeated testing and its consequent, the escalation of Type I error. In the present context, a number of z tests are being performed in the service of a single study. To be more specific, since there are three *basic* parameters among the eight $\lambda^{c\bar{e}}$'s, there are three effect parameters that in the 4×2 configuration are free to vary. Unquestionably, therefore, three independent z tests are possible. Assuming that the study (not a specific comparison) is viewed as the logical unit upon which to base the risk of committing a Type I error, then the performance of three z tests, each at the .05 level, will result in an "experimentwise" error rate in excess of .14. That is, the probability of committing one or more Type I errors will be greater than .14 for the collection of z tests made within the context of the study (see Kennedy, 1978, pp. 165-168, or Myers, 1979, pp. 292-293). And although it can be argued, and convincingly so, that applied researchers in the social sciences traditionally have been overly preoccupied with the harnessing of alpha at the expense of beta, intelligent analysis neverthess requires a healthy respect for the potential promiscuousness of alpha, especially when working with tables that yield sizeable numbers of basic parameters.

One way to exercise control over alpha error is to perform each z test at an enhanced numerical level of significance so that, in the aggregate, the expected number of Type I errors is consistent with the desired collective probability of committing an alpha error. Many may recall that this is the rationale underlying Dunn's test, a multiple comparison procedure frequently used to follow-up omnibus findings in the ANOVA. There is no reason why this same rationale and resultant strategy cannot be used in our work. For example, if it is desired that Type I error be held at .05 in the aggregate, this can be accomplished by performing individual z tests so that E(Type I) = .05. To do this, the desired aggregate alpha is partitioned by the number of z tests on basic parameters that are to be made; then, each z test is executed at the resultant enhanced level of significance. When applied to the normal deviates of Table 6.13, since there are three basic parameters each test in effect will be made at the .017 level of significance (i.e., .05/3 = .017). Thus, to be deemed statistically significant, a z statistic must equal or exceed the critical value of 2.39, the tabled critical value associated with a two-tailed test at the .017 level. Insistence upon this enhanced criterion will ensure that the collective risk of committing an alpha error will be less than .05.

All z statistics presented in Table 6.13 exceed the adjusted critical value of 2.39. It thus follows that the three basic lambdas (and hence the collection of eight lambdas) can be said to have achieved significance at the .05 level, "experimentwise." Bear in mind, however, that the experimentwise approach just advocated is a most conservative approach. In the long run, its use will lead to an excessive commission of Type II errors—particularly when the number of basic parameters being tested is large and the effects being subjected to test have been predicted in advance on the basis of theory or research. In the presence of strong theory, therefore, a slight but justifiable reduction in the Type II error rate can be realized by using an adjusted critical value for one-tailed tests instead of values associated with two-tailed tests. When applied to the tests in Table 6.13, three one-tailed tests each executed at the .017 level would necessitate a z statistic larger than 2.12 (not 2.39) for significance. It is hoped that the future will bring improved strategies for the experimentwise testing of log-linear effects.

Following-up Interactions

In addition to the logit main effects for Variable C, the summary presented in Table 6.12 also revealed a significant chi-square component for Model (8). This observation prompted the conclusion that $\lambda^{b c \bar{e}}$ made a significant contribution ($p < .02$) to the explanation of the residuum about the fitted frequencies produced by the null logit model. Moreover, it can be said that, in addition to main effects for Variable C, there remained differences between T's and F's, but these remaining differences were not uniform over all levels of the response variable. In other words, proportional differences in article preference between T's and F's varied as a function of how they are classified on Variable B, the Intuition (N) versus Sensing (S) variable. Explicating the nature of a first-order logit-model interaction when the logit variable is a dichotomy was a challenge in our earlier discussion of the McLean study. Since the logit variable here consists of four levels, the challenge is somewhat more difficult, but not as difficult as some would make it out to be (e.g., Goodman, 1970, p. 233).

As we saw in the first part of this chapter, the first step in the interpretive process is to martial and summarize relevant tabular information. The first-order-logit effects ($\lambda^{b c \bar{e}}$'s), from the follow-up model selected earlier, and profiles of proportional response to the logit variable made by subjects within combinations in $[BC]$ are basic; thus they are given in Table 6.14. Study the table.

Because the response variable is polytomous, we next identify the level or levels of Variable E that are associated with significant interactive response. The z tests reported in the right-most column of Table 6.14 can assist here. It can be seen that the z's are minuscule for both the NF and the NT articles. Obviously, preference for the inquiry reflected in these two articles is not interactive. This observation is reinforced by the patterns of proportional response where the table shows that approximately 40% of *both* NT's and ST's favor the NT article and about 10% in each of these groupings favor the NF article. Relative to preference to these same two articles, a similar absence of interaction is evident in the NF and SF groups. Thus it follows that our earlier conclusions concerning the direction of main effects for Variable C need not be qualified; it can still be said that proportionately more T's prefer the NT article while more F's favor the NF article. Notice, however, that z tests on effects for

TABLE 6.14. First-Order-Logit Interaction Effects for Variables B and C

Article Preference	NT		ST		NF		SF		
	λ_{11m}^{bce}	p	λ_{12m}^{bce}	p	λ_{21m}^{bce}	p	λ_{22m}^{bce}	p	z test
ST	−.246	.23	.246	.40	.246	.13	−.246	.07	2.05
NF	−.022	.09	.022	.10	.022	.13	−.022	.37	.19
NT	−.027	.39	.027	.38	.027	.09	−.027	.08	.23
SF	.295	.29	−.295	.12	−.295	.35	.295	.49	2.87*
Total	0.0	1.00	0.0	1.00	0.0	1.00	0.0	1.00	

*$p < .05$ (experimentwise).

the *ST* article ($z = \pm 2.05$) and the *SF* article ($z = \pm 2.87$) either approach or attain statistical significance at the .05 level. It is with respect to the choice of the *SF* article, and possibly the *ST* article, that Variables *B* and *C* interact.

To be studied first are the strongest-appearing first-order-logit interaction effects, those associated with the *SF* article. When the 30 *T*'s who chose this article are subdivided into groupings of *NT*'s ($f_{114}^{bce} = 20$) and *ST*'s ($f_{214}^{bce} = 10$), the algebraic signs of corresponding lambdas reveal that a higher proportion of *NT*'s liked the article in question—subsequent, of course, to adjustment for all prior effects in the selected follow-up model. Moreover, of the 64 *F*'s who also happened to select the *SF* article, more *SF*'s ($f_{224}^{bce} = 45$), than *NF*'s (f_{124}^{bce}), chose this article—subsequent to correction for prior model effects. In sum, the overall or main-effects preference of *F*'s for the inquiry embodied in "feeling" articles requires qualification for, as we have just observed, proportionately fewer *NF*'s preferred the *SF* article than did *SF*'s.

Although not as pronounced, interaction also appears to characterize preference for the *ST* article. The algebraic signs of the lambda effects suggest that when the 48 *T*'s who chose this article are additionally classified as *NT*'s ($f_{111}^{bce} = 16$) or *ST*'s ($f_{211}^{bce} = 32$), a greater proportion of the latter seem to have preferred the *ST* article. In point of fact, 40% of *ST*'s chose this article against 23% in the *NT* group. Looking at *F*'s, subjects of this persuasion tended to eschew the *ST* article—only 13 of the 146 subjects classified as an *F* selected it. Even so, differences between *NF*'s and *SF*'s appear to be present. Specifically, more *NF*'s (13%) chose the article than did subjects in the *SF* (7%) classification. These later findings should be considered highly tentative, however, because the *z* tests on the three basic effects ($z = \pm 2.05$) failed to satisfy the more conservative experimentwise criterion established in the previous section.

Interactions can be complex. The experience of having to struggle through the foregoing paragraphs should have convinced most readers that even interactions of the first-order can often be difficult to describe and communicate verbally. Fortunately, description can be clarified and communication enhanced through the use of graphical representations in the delivery of presentations and the preparation of research reports. Recall the graphical representation of binomial logits in Figure 6.1 and how well that graph facilitated our understanding of the first-order-logit interaction that almost achieved significance in the McLean study. The plotting of binomial logits constituted a relatively straightforward procedure in that study because McLean's response variable was a dichotomy. When response is polytomous, as is the case presently, the plotting of logits is more difficult.

When the response variable is polytomous, to construct a graph or graphs similar to Figure 6.1 the researcher must, in effect, find a meaningful way to reduce a polytomy to a dichotomy. The researcher must define one or more sets of conditional odds, translate these odds into binomial logits, and plot the logits. There are several ways that this can be accomplished. Using the interactive outcome of the Peters' study as an example, one approach would be to focus primary interest on the selection of a particular level of the response variable, say the selection of the SF article, and then proceed to calculate the odds of choosing that article over not choosing that article—that is, the odds of selecting SF as opposed to any one of the remaining three articles. Or, the odds of favoring one article as opposed to another specific article may be of interest. To repeat, there are many ways to (a) redefine four levels of response so that response is couched in dichotomous terms, (b) compute conditional odds within the combinations of $[BC]$ as was done previously (see Table 6.10), (c) transform the multiplicative odds to linear binomial logits, as was also done previously (Table 6.10), and (d) plot the resultant binomial logits in a manner parallel to that shown in Figure 6.1. And aside from their salutary communicative value, recall that an advantage of such graphs is that they depict net influences. In the present case, for example, a plot of binomial logits would reflect the aggregate influence of first-order-logit interactions (as defined in Table 6.14), the influence of main-marginal differences in $[E]$, and the influence of logit main effects of Variables B and C.

Much that can be communicated by plots of binomial logits can be communicated equally as well with graphs of within-group proportional response to selected levels of the response variable. The graphs in question are direct, uncomplicated, and well suited for polytomous response variables because the reduction of a polytomy to a dichotomy is not required. In the Peters' study, the construction of such a graph begins with the selection of a level (or levels) of response which, for our purposes, will be the statistically significant ($z = \pm 2.87$) simple interaction between Variables B and C at E_4. That is, for expository simplicity we will graph only the proportion of subjects within intact combinations of Variables B and C who chose the SF article ($f_4^e = 94$). (Of course, comparable graphs could be constructed at other levels of Variable E.) The graph is shown in Figure 6.2.

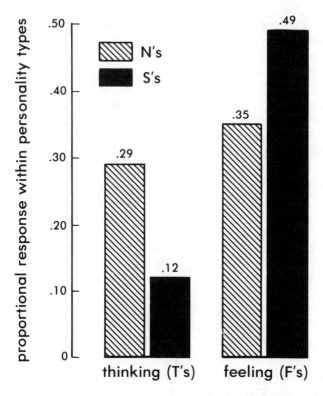

Figure 6.2. Representation of simple first-order-logit interaction between Variables B and C at levels of preference for the SF article.

The manner in which constituent effects contribute to the composite can be gleaned from a study of the figure. In the first place, notice how the significant logit main effects for Variable C, discussed at length on earlier pages, make their appearance in the figure. That is, note that when combined, subjects classified as F's generally preferred the SF article in greater proportions than did either of the groupings of T's. That aside, the pattern of interaction suggested by the $\lambda_{jk4}^{bc\bar{e}}$'s displayed in Table 6.14 can also be discerned in Figure 6.2. Indeed, proportionately more NT's chose the SF article than ST's; but among F's, proportionately fewer NF's than SF's selected the article. But of even greater importance is the fact that the graph suggests how the statistically significant effects, symbolized by the $\lambda_{k4}^{c\bar{e}}$'s and $\lambda_{jk4}^{bc\bar{e}}$'s, act in concert to produce in large part the overall outcome portrayed in the figure. The extent to which each set of effects merits separate analysis, or the extent to which each should be treated synthetically, is a substantive decision best made initially by the researcher—and subsequently by his or her audience in the research community who eventually will pass judgment on the meaningfulness of the work.

7

Additional Applications and Concerns

The contents of this last chapter are varied. Since the adjectives "versatile," "general," and "comprehensive" have been used frequently in this book to describe log-linear methodology, brief discussions of additional applications of this method in support of these adjectives appear to be in order. In the first discussion, one of our working examples will be emended to show that a logit-model analysis need not be limited to one variable of response. Specifically, it will be shown that two or more response variables, called, respectively, two-dimensional logits or multidimensional logits, can be examined simultaneously in a manner loosely analogous to a MANOVA. In a second section we will be reminded of problems inherent with repeated measurements and of a strategy that on occasion can successfully accommodate a repeated measure. The major intent of this section, however, will be to introduce several new models designed to assess symmetrical change over two observational periods. The third new area of application deals with the use of log-linear analogues to causal models that are seen in metric path analysis. It is hoped that these discussions, although brief, are sufficiently ample to convey essential ideas and a direction for further study.

But not all is right with log-linear analysis. As with all new research tools, there are a number of real problems, soft spots, and issues that have yet to be resolved fully. Two of these are of particular concern and have been encountered in this manuscript, although they were allowed to slip by almost unnoticed. In this the final chapter, it is acknowledged that empty cells can cause problems under certain circumstances and, since statistical significance is largely a function of sample size, using only statistical criteria to judge the adequacy of model fit also can be problematic.

We conclude with a discussion that typically one would expect to find in the introductory chapter of a book such as this. The discussion centers about the need for log-linear analysis, particularly the need for analyses that use logit models. The discussion is offered primarily because there are some who would argue—and this writer at one time was among them—that well-established techniques can readily accommodate the work of a log-linear analysis and, therefore, the promotion of log-linear analysis is at best redundant and at worst a wasteful expenditure of time and effort. The discussion has been postponed to the very last so that the arguments advanced in support of the study and use of log-linear methods can be more fully appreciated by the reader.

Thus, the contents of this chapter should give the reader a better idea as to what to pursue next, what problems to avoid in practice, and whether study of log-linear analysis has been a worthwhile investment.

MULTIDIMENSIONAL LOGITS

Let us modify the example that we have been calling the Peters' study so that it can accommodate two response variables. To keep things simple, assume that there are only two explanatory variables. The variables, formerly Variables B and C, will be relabeled Variables A and B. Specifically, the explanatory variables are:

Variable A: Intuition (N) vs. Sensing (S)
   A_1 = subjects classified as N's on the MBTI
   A_2 = subjects classified as S's on the MBTI

Variable B: Thinking (T) vs. Feeling (F)
   B_1 = subjects classified as T's on the MBTI
   B_2 = subjects classified as F's on the MBTI

As before, assume that participants were requested to select one of the four articles reflecting a preferred style of inquiry. (The articles were described in Chapter 6.) But, unlike before, subject preference for a particular article will not be handled as a four-level composite variable. Instead, a particular preference will be jointly classified and entered into a contingency table defined by the crossing of dichotomous Variables C and D. That is, envision the crossing of:

Variable C: Preference for an article with an N component vs. an S component

   C_1 = choice of either the NT or NF article
   C_2 = choice of either the ST or SF article

Variable D: Preference for an article with a T component vs. an F component

   D_1 = choice of either the NT or ST article
   D_2 = choice of either the NF or SF article

The result, of course, is a 2 × 2 table, and a response is a tabulation entered into one (and only one) of the cells of the fourfold *response table*. Moreover, each cell of the response table corresponds to one of the four types of articles.

The intent of the hypothetical study would be to determine whether: (a) the 2 × 2 response *table* is sufficiently different between subjects who have been labeled N's as opposed to S's (i.e., main effects due to Variable A); (b) the response *table* is dissimilar for T's and F's (i.e., main effects for Variable B); (c) differences in resultant *tabular* response between say N's and S's (or T's and F's) are a function of whether subjects are T's or F's (or N's or S's). We recognized the latter as an assessment of first-order interaction between explanatory Variables A and B. By analogy, the situation is like the layout for a two-way factorial ANOVA in which Variables A and B are independent variables, but the dependent "variable" is a 2 × 2 response table. The proposed analysis could serve as a con-servative post hoc assessment of the two response variables in crossed combination to see if it is prudent to examine each response variable separately—just as a MANOVA is sometimes performed to justify the conduct of a series of ANOVAS—or the multidimensional logit analysis could be conducted because effects involving multidimen-sional logits are substantively meaningful in and of themselves. From what we know of the Peters' study, a strong case could be made for the second intent.

Adhering to our decided preference for general logit models (in contrast to models that yield expected odds) and our slight preference for casting models in the multiplicative, the null logit model—Model (0)—for the situation under discussion is

$$F_{ijkm} = \tau \tau^a \tau^b \tau^c \tau^c \tau^{ab} \tau^{ac} \tau^{ad} \tau^{bc} \tau^{bd} \tau^{\overline{cd}} \tau^{abc} \tau^{abd} \qquad (7.1)$$

Model (0) is specified most economically by fitting observed margi-nals [ABC], [ABD], and [CD]. The need to include τ^{ab} in this null model, and hence both τ^a and τ^b, has been discussed at length in connection with the legitimacy of earlier logit models. But there are additional factors (terms) that are also mandatory in this model. With the notable exception of τ^{cd}, notice that all interactions that do *not* portray logit Variables C and D *acting jointly* are also built into the model. These six interactive parameters estimate effects that are not totally relevant to the desired response, a response that is the composition of a fourfold table completely defined by the crossing of C and D. The six parameters, therefore, must be incorporated into this and other logit models to effect control over these extraneous effects. The parameter τ^{cd} also belongs in this and other models, but for a different reason. If the interaction symbolized by τ^{cd} should turn out to be prominent, this would mean that subsequent to the fitting of a model, differences are manifest among the four geometric means in the cells of the [CD] configuration, even after corrections are made for main effects due to C and D. But are we really concerned with differences among cell *frequencies* within the response tables? No. Strictly speaking, our concern is limited to the *expected proportions* within the four cells of the response table and whether such tables as this are similar over levels of Variables A, B, etc. Thus, just as we earlier put τ^e into the unidimensional null logit model for this study (see Table 6.11), extending our logic will have us put τ^{cd} in the null logit model for the current version of the study.[1]

Presented in hierarchical fashion in Table 7.1 are models for a two-dimensional logit analysis where crossed Variables C and D define response. Model (0) is given by Eq. (7.1) and, as we have noted, it does not contain a second-order or third-order interaction that simultaneously involves C and D. Model (1) contains the 13 parameters seen in Model (0) but in addition has a fourteenth parameter, the second-order interaction τ^{acd}. Model (2) possesses the 14 parameters of Model (1) plus τ^{bcd}. Finally, the addition of third-order τ^{abcd} to the aforementioned factors gives the saturated model. When fitted by a computer program such as BMDP/4F or ECTA, residual L^2's with corresponding ν's are outputted for each model. The researcher compiles respective residual L^2's and with a pocket

---

[1] Should the null logit model be accepted and should it also be found that for all practical purposes $\tau^{cd} = 0$, then not only can it be said that there are no differences in proportional response within [CD] over levels of A, B, etc., but differences do not even exist if frequencies are inserted in the cells of the response table.

TABLE 7.1. Two-dimensional Logit Models where Variables A and B are Explanatory

No.	Logit Model	Marginals Fitted
(0)	Null Logit	$ABC, ABD, \overline{CD}$.
(1)	Due to Variable A	$ABC, ABD, A\overline{CD}$.
(2)	Due to Variable B, Given Variable A	$ABC, ABD, A\overline{CD}, B\overline{CD}$.
(3)	Due to Interaction	$AB\overline{CD}$.

Note. Crossed Variables C and D define the logit response table.

calculator computes component L^2's and subjects them to test as we have done on numerous occasions in the past.

Judging effects to be significant also proceeds as before. Typically, the residual L^2 for the null is examined first because it is a composite indicator of the residuum about the fit of the null. If the composite residual is significant, further testing is clearly warranted. Relative to further tests, suppose the component chi-square for Model (1) was shown to be large and significant. This would indicate that τ^{acd} is making a significant contribution to the reduction, and hence explanation, of the residuum from expectancies given by the null-logit model. Such a reduction points to significant main effects due to Variable A. In other words, the standardized-to-proportional-unity table of response for subjects classified as N's (i.e., $\Sigma_k \Sigma_m P^{acd}_{1km} = 1.00$) is in some significant way different from the corresponding response table produced by S's.

Tabular differences, of course, would need to be explored. A first step in this exploration would be a study, at each level of A, of the algebraic signs and distribution pattern of the τ^{acd} effects belonging to the model selected for follow-up, and a study of subsequent z tests on the logs of these effects. Following explication of the multidimensional result, if it should be of interest, a series of unidimensional logit model analyses can be conducted along the lines by which multiple ANOVA's are sometimes used to follow-up a MANOVA. Because the composite multidimensional test tends to be conservative, successive unidimensional models can be examined with less fear of probability pyramiding.

Should the component L^2 belonging to Model (2) be judged to be significant, main effects due to Variable B are indicated. Because τ^{bcd} comes after τ^{acd} in the model, main effects for B are evident over and above those that might be due to Variable A. Finally, if significance is observed for τ^{abcd} in the saturated model, then in addition to main effects which might have been detected, an interactive pattern in response tables is present. This interaction can be approached in several ways. For example, by plotting the τ^{abcd}'s, it might be of interest to describe first differences in the 2×2 response tables between T's and F's who are also N's, and then differences in tabular response between T's and F's who are S's. The stronger the interaction, the greater the discrepancy to expect between these two descriptions. With a little ingenuity, the graphical procedures used to display earlier interactions can be extended to describe and communicate the interaction in question.

Within reason, general log-linear models can be constructed to accommodate any number of explanatory variables and response variables. As an example, imagine a four-dimensional contingency table in which only one variable, say Variable A, is considered explanatory. Rather than performing a series of unidimensional logit model analyses, a sagacious initial analysis would be one in which Variables B, C, and D are crossed to form a three-dimensional response table. The analysis would in some ways resemble a one-way ANOVA and would require the fitting of only two models. The first, the null logit model, would contain 16 general parameters and would fit observed configurations $[ABC]$, $[ABD]$, $[ACD]$, and $[BCD]$. The saturated model would constitute the second model. Acceptance of the latter, which acknowledges the need to fit τ^{abcd}, can be taken to mean that there is a difference in the three-dimensional response table between at least two levels of Variable A. Examination of the τ^{abcd}'s at different levels of A will point to the location of the difference or differences. Following the multidimensional analysis, separate unidimensional logit-model analyses can be undertaken with greater insight and respect for alpha.

REPEATED MEASUREMENTS AND CHANGE

Repeated measurement variables, common to ANOVA designs, are encountered occasionally when working with qualitative variables. And as a rule, when multiple observations are made on the same subjects, the assumption of response independence, basic to both the F and multinomial distributions, is contradicted. For the ANOVA, contradiction of the independence assumption leads to a positively biased F test, whereas violation of this assumption for the chi-square approximation to the multinomial leads to the retention of null hypotheses, at a given level of significance, more often than it should (Maxwell, 1961, p. 26). Over the years a number of strategies have been developed for the ANOVA that, for the most part, have overcome difficulties associated with repeated measures. Unfortunately, similar comprehensive strategies have yet to be developed for the log-linear. There are, however, several less direct approaches that on occasion can be used both to detect change and to describe the direction of that change.

Approaches to Problems of Repeated Measures

It may be recalled (Chapter 5) that Holton and Nott, in their study on reflective teaching, encountered a common repeated measurements situation. They wanted to measure change in the complexity of written response both prior and subsequent to an instructional intervention. Before experiencing reflective teaching or control group activities, the writings of subjects were classified into one of four response modes: analytical, evidential, declarative, or indeterminate. Following three weeks of intervention, written responses were obtained again from the same subjects and again subjected to the above classifications.

Experience with the ANOVA might tempt an investigator to structure a two-level pre-/posttest variable and to cross that variable with the four-level mode of response variable. The result would be a 4×2 table. For a moment, let mode of response be Variable A and pre-/posttest be Variable B. Now, if there were little or no change in the way written specimens were classified from pretest (B_1) to posttest (B_2), one consequent would be that $f_{i1} = f_{i2}$ for all levels of A, and the model that would fit these data well would be $F_{ij} = \tau \tau_i^a$. If change did occur, and if the result of change was that $f_{i1} \neq f_{i2}$ for two or more levels of A, then the model would become $F_{ij} = \tau \tau_i^a \tau_{ij}^{ab}$. But, unfortunately, this appealing analysis is flawed because by counting a subject twice in the table, the independence assumption (discussed in Chapter 4) is most likely being violated.

Recall that Holton and Nott avoided the problem by crossing the four-level pretest variable (Variable A) with its replicate, the posttest variable (Variable C). The outcome was a 4×4 table in which subjects were counted but once. Had there been three or four testing periods, this approach would likely be intractable. Also, the degree of correlation between the crossed variables determines the feasibility of this approach. For example, had there been little change between testings, hence a high correlation, the approach used by Holton and Nott would not be advisable because most frequencies would be found in cells on or near the principal diagonal, leaving off-diagonal cells either empty or with few frequencies. Admittedly, as we will learn later in this chapter, a few zero cell frequencies do not constitute a major problem if they are related to sampling procedures (i.e., sampling zeros). Holton and Nott had an empty cell in their 4 X 2 X 4 table (see Table 5.7), yet the overall distribution of frequencies did not, in their view, militate against the approach in question. In fact, if permissible, the approach has much to recommend it. It is direct, it lends itself to straightforward applications of log-linear models, and as we saw in their study it permitted Holton and Nott to test for treatment effects on posttest response subsequent to adjusting for initial pretest effects.

Other ways to measure change have been proposed for situations in which a group is tested or observed on two separate occasions. We are referring here to models used to describe *symmetry* and *quasi-symmetry*. Suppose, for example, that it is evident that changes have occurred in classifications made at two points in time. This may prompt a question concerning the nature or direction of this change. Change in the table may have been random and hence diffuse, or systematic and channeled in one direction. Before pursuing this matter further, let us set up a one-group pre-/posttest design (called a *panel study* in sociology) using the familiar data from Holton and Nott, so that we can use these data to explore some of the problems of trying to detect change in such a design.

Of data reported in Table 5.7, we will consider only those belonging to one group, the group of 55 subjects who participated in the reflective teaching treatment. With respect to this group, we want to know if significant change in mode of responding took place between the pretest and posttest. As mentioned, a *direct* assessment of this central question appears to be beyond the reach of log-linear methods at this time. To appreciate the problem better, three hypothetical data sets have been created which preserve values in [A] for the 55 treated subjects in the Holton-Nott study, as shown in Table 7.2.

In Table 7.2, the data on the left where all frequencies are found on the *principal diagonal* depicts absolutely no change from pretest to posttest. Change is seen in the middle set, however. Note that the change is distributed somewhat uniformly throughout the table; that is, students assigned to a particular pretest category tend to distribute themselves evenly over levels of the posttest variable.

TABLE 7.2. Hypothetical Data Sets for the Holton-Nott Study

Pre-test	Posttest				Posttest				Posttest			
	a	*e*	*d*	*i*	*a*	*e*	*d*	*i*	*a*	*e*	*d*	*i*
a	15				4	5	2	4				15
e		17			5	5	1	6			17	
d			6		2	1	2	1		6		
i				17	4	6	1	6	17			
	No Change				Uniform Change				Extreme Change			

On the right we see the opposite of that shown to the left, namely dramatic change. To the extent to which we give ourselves permission to regard the testing variable as an *ordered* variable, a negative correlation between testing variables can be said to exist in the data to the right.

At first thought, it would seem that conventional log-linear models such as those summarized in Table 4.1 could be used effectively to detect departure from the pattern shown by data on the left. After all, such departure would be indicative of change between testings. But, the near-perfect fit of *any* model for two-way tables, even the one-parameter model, would indicate deviation from the no-change pattern on the left. The independence model in particular would appear at first thought to be most informative, for in the absence of change the pretest and posttest variable would be highly correlated and the independence model would not fit data well. Consequently the saturated model would be accepted. But will a good fit be realized for the independence model if it is applied to the extreme-change data on the right? The residual chi-square for this model when applied to the extreme-change data turns out to be $L^2(4) = 168.88$, $p < .00$. In short, for both situations we would have no choice but to accept the saturated model and its interpretation that the pretest and posttest variables are correlated, yet in the latter situation change *has* transpired. Finally, to make matters even more confusing, the independence model fits the uniform-change data very well [$L^2(9) = 4.47$, $p < .85$], but again change is indicated. To cut a long story short, a simple set of rules for the detection of change using log-linear models in the present context does not appear to exist.

Log-Linear Models for Symmetry

If it can be assumed, however, that change has taken place, as evidenced by ample frequencies in cells off the principal diagonal, one can ask whether the change in one direction is the same as the change in the other direction. If so, the "quantity" of change given by nonzero entries in off-diagonal cells located in the upper-right portion of the square table will be mirrored in the lower-left portion of the table. Put differently, to the extent to which change takes place, frequencies will be found in cells that do not reside on the principal diagonal, and to the extent to which $f_{ij} = f_{ji}$, when $i \neq j$, change is balanced, bilateral, and said to be symmetrical. And if change can be described as symmetrical, then it cannot be said that the state of affairs at the second testing is different from that observed at the first testing. Relative to Holton-Nott, where it was hypothesized that exposure to reflective teaching would result in students being more *analytic* in their posttest performance, symmetry would militate against support for their hypothesis. Rejection of symmetrical change, however, could be taken as first evidence in support of the hypothesis, for rejection would suggest that posttest performance is different from performance on the pretest. Rejection, therefore, would justify further study into the nature of the change.

We will subject to test the hypothesis of symmetrical change using data obtained on the 55 treated subjects in the study by Holton and Nott. But first find the 4 × 4 table on the left of Table 7.3 which contains observed frequencies organized by levels of the pretesting variable (Variable A) and the relabeled posttest variable (Variable B). Because we are only concerned with the amount and nature of change from pretest to posttest, ignore from now on the cell entries in the principal diagonal. Now if the change that transpired had been symmetric, what would we expect to see in the off-diagonal cells to the upper-right and lower-left? The expected outcome is also presented in Table 7.3. The expected frequencies may be obtained in one of two ways. One way is to calculate them directly by averaging observed frequencies in corresponding off-diagonal cells. That is, for cells not on the principal diagonal,

$$F_{ij} = F_{ji} = (f_{ij} + f_{ji})/2 \tag{7.2}$$

Or, the F_{ij}'s for expected symmetry can be given by a log-linear model that will yield F_{ij}'s identical to those computed directly with Eq. (7.2).

TABLE 7.3. Observed Frequencies and Those Expected under the Hypothesis of Symmetrical Change

Pre-test	Posttest					Posttest				
	a	*e*	*d*	*i*	[*A*]	*a*	*e*	*d*	*i*	[*A*]
a	8	2	1	4	15	8	4	1	6	19
e	6	4	2	5	17	4	4	2	4.5	14.5
d	1	2	2	1	6	1	2	2	1	6
i	8	4	1	4	17	6	4.5	1	4	15.5
[*B*]	23	12	6	14	55	19	14.5	6	15.5	55
	Observed Frequencies					Symmetric Frequencies				

Before we formulate and fit a log-linear model for symmetry, preparatory comments concerning expected main marginals, chi-square, and degrees of freedom are in order. First, examine the main marginals of the 4 × 4 table of expected frequencies in Table 7.3 or, for that matter, any $k \times k$ table that satisfies the criterion of symmetry. Note that under symmetry, $F_i^a = F_j^b$ when $i = j$. In words: symmetry will produce *homogeneous marginal distributions*. By the same token, if observed marginals are not homogeneous, then to the extent to which they are not, the potential of achieving symmetry is reduced. Second, a comparison of a $k \times k$ table of observed frequencies with a conformable table of frequencies to be expected under symmetry can be routinely accomplished through the use of a goodness-of-fit chi-square. But, since the cells on the principal diagonal are not involved in our thinking, they should not be involved in the tabular comparison. Consequently, the L^2 or χ^2 is computed using only $k(k-1)$ cells. Since $k = 4$ in our example, only 12 cells will participate in the calculation. For the example, the likelihood-ratio statistic is

$$L^2 = 2 \sum_{i \neq j} (f_{ij})[\ln (f_{ij}/F_{ij})]$$

$$= 2[\ln (2/4) + \cdots + \ln (1/1)]$$

$$\doteq 3.56$$

Finally, the number of degrees of freedom also needs to be modified. Granted, $k(k-1)$ cells are involved in the computation of L^2, but since one complete side of the table (e.g., the upper right) can be constructed from knowledge of the other side, only one-half of these cells (i.e., 6) were free to vary. Hence, in general, $\nu = k(k-1)/2$. In particular, our result was $L^2(6) = 3.56$, $p < .70$. The hypothesis of symmetrical change cannot be rejected.

We are most interested, however, in testing for symmetrical change using log-linear models, for even though the direct and the log-linear approach will yield the same results, for this and related problems the log-linear is more general and generative. A log-linear approach involves: (a) expanding a two-dimensional $k \times k$ table into a three-dimensional $(k-1) \times (k-1) \times 2$ situation, (b) specifying a model that, if acceptable, describes symmetry, (c) fitting the F_{ijk}'s given by the model to observed f_{ijk}'s, many of which turn out to be *structural zeros*, and (d) comparing the fit of the specified model to another model or models to judge its acceptability. Again, the data provided by Holton and Nott will be used to demonstrate the approach just outlined.

Expanding to Three Dimensions. For the tables shown in Table 7.3 it has been established that the row variable is Variable A and the column variable is Variable B. We desire to introduce into this situation a new dichotomous variable, Variable C, where C_1 consists of off-diagonal entries in the lower-left "triangle" of a two-dimensional table and C_2 consists of the frequencies in the upper-right triangle of a two-dimensional table. Variable C, therefore, partitions the two-dimensional tables shown earlier into two triangular tables that, subsequent to a reorganization, can be compared. It is hoped that the contents of Table 7.4 convey the reorganization and possibility for a tabular comparison.

Consider first the observed frequencies shown in the upper portion of Table 7.4. Specifically, notice how the *reduced* table of observed frequencies (i.e., the table of observed frequencies without entries in the principal diagonal) has been partitioned into a lower and upper triangular matrix. Then to make the tables commensurate, the upper table is turned about to bring it into juxtaposition with the lower triangular matrix. To complete each, *structural zeros* (to be discussed in a latter section) are inserted in the void cells. In sum, for observed frequencies we now see a $3 \times 3 \times 2$ contingency table where Variable A is the new three-level row variable, B is the three-level column variable, and C is the two-level partitioning variable.

Consider next the frequencies in the lower portion of Table 7.4. We recognize these to be the frequencies under the hypothesis of symmetry. Using parallel procedures, the two-dimensional table of

TABLE 7.4. Observed Frequencies and Frequencies Expected under the Hypothesis of Symmetrical Change

	Observed Frequencies											
	Posttest				Posttest				Pretest			
Pre-test	*a*	*e*	*d*	*i*	*a*	*e*	*d*	*i*	*a*	*e*	*d*	*i*
a						2	1	4				
e	6						2	5	2			
d	1	2						1	1	2		
i	8	4	1						4	5	1	
	(1) Lower Triangle				(2) Upper Triangle				(3) Transposed Triangle			

	Expected Frequencies											
a												
e	4								4			
d	1	2							1	2		
i	6	4.5	1						6	4.5	1	
	(1) Lower Triangle								(3) Transposed Upper Triangle			

Note. In the log-linear analysis of these $3 \times 3 \times 2$ tables, structural zeros are entered in cells that are shown to be empty.

expected frequencies can be partitioned so as to yield a $3 \times 3 \times 2$ contingency table. The idea, of course, is to compare the observed and expected frequencies within the three-dimensional context; if it should be found that expected frequencies approach the observed, symmetrical change is indicated.

The Model for Symmetry. Our present task is to deduce the factors (or terms) that will appear in a log-linear model that will produce the expected cell frequencies seen in Table 7.4. Clearly, differences among the F_i^a's and the F_j^b's are to be expected; therefore, τ_i^a and τ_j^b belong in the model for symmetry. But since $F_1^c = F_2^c$ by our construction of symmetry, τ_k^c will not be found in the model. Then if we collapse over levels of C to view expected frequencies in $[AB]$, it will become clear that the F_{ij}^{ab}'s cannot be generated by using only information in [A] and [B]. Thus, it follows that interaction between A and B also must be acknowledged in the model. If, however, first-order interaction is present between Variables A and C, which would mean that effects for A are not the same at both C_1 and C_2, symmetry is obviated. Also, for the same reason, to have symmetry there cannot be interaction between Variables B and C. Finally, should τ^{abc} be needed in the model, suggesting a differential pattern of interaction between Variables A and B over levels of C, change cannot be symmetric. We conclude, therefore, that the model representing symmetrical change is,

$$F_{ijk} = \tau \tau_i^a \tau_j^b \tau_{ij}^{ab} \tag{7.3}$$

Fitting Expected Frequencies. The four-parameter model above will generate the F_{ijk}'s shown in Table 7.4. To do this, our hand calculations or our computer's iterative fitting algorithm must be constrained so that nonzero estimates are not given for the cells in which we inserted zeros. Both ECTA and the newer 4F version of BMDP permit these designated cells to be excluded from the iterative fitting process.

Assessing Goodness-of-Fit. A residual L^2 may be used to determine how well the F_{ijk}'s produced by Eq. (7.3) fit observed data. Keep in mind, however, that structural zeros are to be found in the three-dimensional tables of observed expected frequencies. When doing the analysis by hand, these cells are simply excluded from consideration, for ultimately we would be asked to divide a zero by a zero, and division by zero is not a legitimate arithmetic operation. Strictly speaking, we are arbitrarily defining $0/0 = 0$. Also, when determining the number of degrees of freedom, we must subtract from the total a degree of freedom for every structural zero. Moreover, remember that half of the nonzero cells have F_{ijk}'s that are determined by the other half. If you are given one triangular table, you can generate the F_{ijk}'s in the other table. Thus, in general, $\nu = k(k-1)/2$. For our example, $\nu = (4 \times 3)/2 = 6$.

Finally, be reminded that we know how well Eq. (7.3) fits observed data, for we have computed the L^2 earlier in connection with the two-dimensional approach to the problem. Recall that $L^2(6) = 3.56$, $p < .70$. At this point we tentatively conclude that the symmetrical model constitutes a good fit, and therefore it is doubtful whether those who changed from pretest to posttest did so in the direction of being more analytical.

Proportional Symmetry.   Envision a situation where there are more tabulations in one triangular table than in another, i.e., $f_1^c \neq f_2^c$. Even though symmetry, as previously defined, cannot come about in this situation—for one thing, we would have heterogeneity of the distributions in [A] and [B]—nevertheless, the pattern of frequencies in the upper and lower triangles could be symmetric to a proportionality constant. For example, if there were twice as many frequencies in the lower triangle, and if symmetry could be produced by dividing each cell frequency in the lower triangle by two, a modified form of symmetry, *proportional symmetry*, is evidenced. Notice in the Holton-Nott data that tabulations in the lower triangular table exceed those in the upper table. Specifically, $f_1^c = 22$ while $f_2^c = 15$. Proportional symmetry would exist in the off-diagonal cells of the original two-dimensional table if

$$f_{ij} = (f_2^c/f_1^c)f_{ji} \qquad (7.4)$$

Thus, accepting that there are differences in [C], cell frequencies to be expected in the $3 \times 3 \times 2$ table under the hypothesis of proportional symmetry are obtainable by fitting [C] in addition to [AB]. The model, therefore, that describes this modified form of symmetry is

$$F_{ijk} = \tau \tau^a \tau^b \tau^c \tau^{ab} \qquad (7.5)$$

To the nearest tenth decimal place, this model yields the following cell entries:

4.8			3.2		
1.2	2.4		.8	1.6	
7.1	5.4	1.2	4.9	3.7	.8
Lower Triangle			Upper Triangle		

It can be seen that the relations given by Eq. (7.4) do in fact apply to these data. Consider, for example, the expected frequency for the second row and second column of the original two-dimensional table. We have

$$F_{21} = (f_2^c/f_1^c)F_{12}$$
$$= (22/15)(3.2) = 4.7$$

where a slight discrepancy due to rounding error is seen between the result (4.7) and F_{21} in the table above (i.e., 4.8).

What implications are associated with acceptance of the model for proportional symmetry? First, acceptance indicates symmetry to a proportionality constant, as has been illustrated. Second, since more people fall in one of the triangular tables, there is more change flowing in one direction than in the other. To aid in the explanation of this point, assume for the moment that the principal variable in Holton-Nott was a true ordered polytomy. The lower end of the ordinal scale underlying the variable was occupied by analytical statements, while the upper end was defined by indeterminate statements. If that were so, acceptance of Eq. (7.5) would mean that of the 37 subjects who changed from pretest to posttest, more of them changed toward the lower end of the variable scale—toward the analytic—than changed in direction toward the indeterminate class at the upper end of the scale. In fact, 22 subjects, were judged to have a "lower" classification on the posttest, whereas only 15 were placed in a higher category on the posttest. Acceptance of the proportional change model, with the directionality of change just discussed, would speak in favor of the investigators' substantive hypothesis that reflective teaching can promote change toward higher level response, particularly analytic response. Acceptance, of course, would be most meaningful if the principal variable were, in fact, ordered—not partially ordered as was the case. Even so, as we shall soon learn, one should accept and hence interpret this model only if done so with knowledge of the performance of other models, e.g., symmetry and quasi-symmetry.

Before pointing out some limitations of this model, let us record how well it fits the data gathered by Holton and Nott. The resultant chi-square would be found to be $L^2(5) = 2.32$, $p < .80$. Hence, the fit is slightly better than that provided by Eq. (7.3), the model for symmetry. Note that the number of degrees of freedom is one less than for symmetry. To be more specific, $\nu = [k(k-1)/2] - 1$, because the model for proportional symmetry contains τ^c and, as a result, there will be one less expected cell frequency that can vary.

As mentioned, despite its appeal the model in question can be misleading if interpreted in isolation, for it makes no adjustment for differences among levels of the initial variable, the pretest variable. For example, by consulting the two-dimensional table of observed frequencies in Table 7.3, then summing frequencies in cells off the principal diagonal, the number of subjects in each pretest category who changed categories during the experiment can be identified.

When these subjects are so classified, the distribution is not uniform. Using $\bar{f}_i^a$ to denote the number of subjects in the ith level of pretest *who changed*, we find that $\bar{f}_1^a = 7$, $\bar{f}_2^a = 13$, $\bar{f}_3^a = 4$, and $\bar{f}_4^a = 13$. Moreover, because there are observed inequalities in levels of pretest, we can expect to see inequalities by level of pretest in the *expected* frequencies given by both the model for symmetry and the model for proportional symmetry. It is with the latter model, though, that this is a concern, for acceptance of proportional symmetry can lead, as we have learned, to the conclusion that a greater number of subjects changed toward a particular end of an ordered scale. The problem is that such a conclusion may be due to the simple fact that there were more subjects at some levels of the pretest than others, thus permitting more change to occur in a given direction.

Consider an extreme example where we will suppose that about half of the subjects were put into A_4 during pretest and, as one would normally expect, the number of subjects in this pretest group who changed far exceeded that in any other pretest group. In what direction can these subjects change? Since change can only take place toward the lower end of the scale, and since there are so many subjects exhibiting this manner of change, the stage is set for possible acceptance of proportional symmetry and advancement of the finding that change flows toward the lower end of the scale. Granted, more subjects may be changing toward the lower end; but if pretest standing is taken into account, are *proportionately* more subjects at A_4 moving toward the lower end than subjects at say A_1 who are moving toward the higher end of the scale?

For a less extreme example, consider the expected frequencies under proportional symmetry for the Holton-Nott data and compare the number of subjects manifesting change at respective levels of the pretest. These expected frequencies are $\overline{F}_1^a = 8.9$, $\overline{F}_2^a = 10.0$, $\overline{F}_3^a = 4.4$, and $\overline{F}_4^a = 13.9$. There are more expected subjects at A_4 who will move lower on the variable scale than expected subjects at A_1 who will distribute themselves over higher levels on the posttest scale. Again, if adjustments are made for the differences in pretest standing, would change be symmetrical or, instead, would change flow in a particular direction or toward a particular level of the posttest variable? To answer this and the question of the preceding paragraph, a model for quasi-symmetry needs to be built and tested.

Quasi-Symmetry. A model for quasi-symmetry, in addition to all prior adjustments made for symmetry and proportional symmetry, must fit the observed marginals for the pretest variable in the three-dimensional situation, a $3 \times 3 \times 2$ situation at present. But since the pretest is the row variable for the lower triangular table at C_1 and the column variable for the upper table at C_2, to fit pretest marginals we need to fit the observed configurations for both $[AC]$ and $[BC]$. By fitting both, however, adjustments will be made also for inequalities in the number of subjects who change when classified by level of posttest performance. Quasi-symmetry, therefore, is represented by expected frequencies from the full first-order model, namely

$$F_{ijk} = \tau \tau^a \tau^b \tau^c \tau^{ab} \tau^{ac} \tau^{bc} \tag{7.6}$$

If Eq. (7.6) gives a good fit, it means that change is symmetrical following adjustments for unequal marginals. For Holton and Nott, a good fit would mean that change is *not* flowing primarily toward the lower end of the scale where analytic response resides. Should the model for quasi-symmetry not fit well, the saturated model would be adopted and change would be described either as a shift toward one end of the scale, if the scale is ordered, or as effects whose nature will be revealed through a study of the τ^{abc}'s.

Expected frequencies given by the quasi-symmetrical model for the working data are shown below:

5.7			2.3		
1.3	1.7		.7	2.3	
8.0	4.0	1.0	4.0	5.0	1.0
Lower Triangle			Upper Triangle		

Comparing the above to observed frequencies in Table 7.3 gives $L^2(3) = .33$, $p < .98$, where $\nu = [(k-1)(k-2)]/2$. The model obviously fits well. Should we accept it, our conclusion will be that change is symmetric, not directed toward the analytic, if differences in main marginals (e.g., differences in pretest tabulations) are taken into account.[2]

---

[2] Since the L_s^2 produced by the fit of the symmetry model represents residual deviation from both symmetry and homogeneity of marginal distributions, but the L_{qs}^2 produced by the model for quasi-symmetry represents only deviation from symmetry, a test on the hypothesis of homogeneity of marginal distributions is provided by $L_{hm}^2 = L_s^2 - L_{qs}^2$, where ν for L_{hm}^2 is obtained by parallel subtraction. In the present example, $L_{hm}^2(3) = 3.56 - .33 = 3.32$, which falls short of significance at traditional levels. The marginals cannot be said to differ from pretest to posttest.

There is still another view of quasi-symmetry. Consider the two-dimensional table in Table 7.3. By fitting Eq. (7.6), the observed main marginals of the two-dimensional table are also fitted. Recall that to achieve symmetry, the marginals of the two-dimensional table first have to be homogeneous (i.e., $f_i^a = f_j^b$). If they are not, the capacity to achieve symmetry is accordingly reduced and the L^2 for the simple symmetry model becomes large. By fitting $[A]$ and $[B]$ in the two-dimensional configuration, the model for quasi-symmetry in effect acknowledges the possibility of heterogeneity of marginal proportions and, within the limits of this acknowledged constraint, proceeds to generate expected cell frequencies that are as symmetric as they can be under the circumstances. Hence, acceptance of Eq. (7.6) can be taken to mean that symmetrical change has taken place given the fact that pre-/posttest marginals were not homogeneous for the two-dimensional table, or were not equal in the three-dimensional situation.

Concluding Remarks. Before we summarize our investigation of change (or lack of change) for the 55 treated subjects in the Holton-Nott study, several points should be made. The first is that questions that we have asked of data in this chapter were not the same as those asked by Holton and Nott in their analysis described in Chapter 5. Second, recall that even the findings in support of change advanced in Chapter 5 were somewhat tentative due to the fact that the composite logit-model test failed to achieve significance at a traditional level (see Table 5.12). With these points in mind, let us examine a summary of our work given in Table 7.5.

TABLE 7.5. Summary of Analyses of Change of Treated Subjects in the Holton-Nott Study

Model No.	Model	Residual			Component		
		L^2	df	p	L^2	df	p
(7.3)	Symmetry	3.56	6	.75			
(7.5)	Proportional Symmetry	2.32	5	.85	1.24	1	.25
(7.6)	Quasi-symmetry	.33	3	.95	1.99	2	.40
(Sat.)	Saturated	0.00	0		.33	3	.95

Adhering to past practice, an examination of residual chi-squares indicates that none of the models in Table 7.5 can initially be excluded due to gross poorness-of-fit. Next, an examination of component chi-squares reveals that the saturated model can be passed over in favor of an unsaturated model. In so doing, the hypothesis that pre-/posttest change would be in the direction of analytic statements will not find support in this analysis. In short, most analysts would select Model (7.3), the symmetry model, since it is most parsimonious yet consistent with observed data. Thus, there was some degree of change in classifications but whereas some subjects moved toward the analytic, others moved away from the analytic. As so often happens in research, one type of question pursued by its logical method will produce from data one result while another question, pursued by its method, will yield a different, sometimes contradictory, result.

CAUSAL MODEL ANALOGUES

Conducting research from which one is able to infer causality is the ultimate goal of science. Conditions necessary and sufficient to advance causal inferences, however, are understandably complex and are best left to other sources for their exposition (e.g., Cook & Campbell, 1979; Heise, 1975). Suffice it to say that the requisites for serious causal thinking in the study of social behavior are rarely approached in the absence of well-controlled experimentation or a program of descriptive research based on extremely strong theory. With this caveat in mind, it is with circumspection that we devote the next few pages to a brief introduction to the use of log-linear models in path analysis and its application to causal thinking.

Modern path analysis subsumes a number of techniques that when used appropriately are useful for the study of the direct and indirect effects exerted by some variables on other variables. The earliest principles were formulated in the 1920s by Sewall Wright, a biologist (Heise, 1975, p. 112). The last two decades have seen extensive refinement and use of path techniques particularly in fields like sociology, economics, and political science where experiments, the classical vehicle for the study of causality, are difficult if not impossible to conduct. However, more recent developments have led to path analytical techniques being extended, integrated, and in a sense even superseded by a more comprehensive approach to modeling called *linear structural equations* or the *analysis of covariance structures* (Jöreskog, 1978).[3] Unlike log-linear models, structural equation modeling is done with either observed or latent variables (constructs inferred from observable variables) where observed variables or indicators are measured on interval scales and least-squares linear regression serves as the basis for the modeling system. Nevertheless, structural equations and log-linear models have much in common and both share the same basic approach— to create a series of plausible models to explain relations or effects, to estimate parameters in these competing models, to determine the goodness-of-fit of competing models through the use of the chi-square, and finally, to choose a model that represents the simplest, yet most plausible, explanation that appears to be consistent with observed data.

Developing a Causal Scheme

Returning to older forms and assuming of the reader an elementary background in conventional path analysis, we will now illustrate how a number of path analytical procedures can be applied in a qualitative setting. To fix ideas, let us review several basic features that are common to path analyses irrespective of setting. First and foremost, a sound system of a priori hypotheses (i.e., theory) is central. Causal path analysis is used best as a rifle where careful aim is taken to test a theory; it is an abuse to use it as a shotgun to scavenge for models that simply appear interesting. Sampling is the second important element. The sampling plan must support each causal prediction that becomes part of the model. Shortly we will use again the data gathered by McLean to test a model that we will develop, but to do so we will find it necessary to take liberties and assume sampling procedures that in fact were not executed by McLean.

From the theory one or more *path diagrams* are created which depict how important variables are interrelated. For our purposes, we will consider only the case where models depicted in diagrams are understood to be *recursive*, i.e., all causal influences flow in one

---

[3] Whereas Karl G. Jöreskog is generally regarded as the major developer of the structural equation movement, his writings, like the writings of many pioneers in a field, tend to be overly technical, particularly for substantive researchers. For a less technical overview of this emerging field, an exposition and review of the literature by Bentler (1980) is recommended.

direction, from left to right, across the diagram. As is standard, arrows will be used to connect variables that are related. The absence of an arrow between two variables indicates that the variables are believed not to be related. If a pair of variables is connected by a two-headed arrow (↔), a bilateral association is indicated. If the arrow has one head, coming from one variable and pointing toward another, a causal linkage in the suggested direction is posited. The kind of diagram described here is shown in Figure 7.1.

For a concrete example, let us return to the study conducted by James McLean and modify it as need be to suit our pedagogical purposes. Turning to the hypothetical scenario in Figure 7.1, note that it begins on the left with two *exogenous* variables: Variable *A* (high vs. low ability) and Variable *C* (black vs. white students). Exogenous variables have "causes" outside the diagrammed system and are easily identified because they have no arrows pointing at them. If our theory held that exogenous Variables *A* and *C* were related, they would be connected by a curved line with arrowheads at each end. To test for the hypothesized relationship, a sample representative of the population of inference would need to be obtained and then jointly classified by Variables *A* and *C*. This cannot be said to have happened in the McLean study, but for a preliminary stage of our analysis we will assume momentarily that it has. Notice too that exogenous Variable *A* has only one *direct* arrow or path from it to *endogenous* Variable *D*. Variable *D* (graduation vs. nongraduation) is the ultimate response variable in the diagram and can be called endogenous because it is affected by

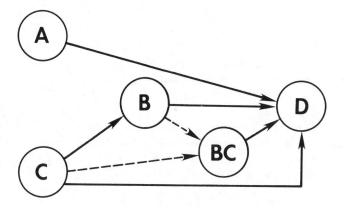

Figure 7.1. Path diagram showing four main variables and an interactive variable to be used in conjunction with the McLean study.

other variables in the system, i.e., it has one or more arrows pointing at it. Variable D is also receiving the pointed end of a direct arrow from Variable C. Variable C also is said to have a *direct* effect on students' choice of university, penultimate endogenous Variable B. Again, to legitimately subject this predicted effect to formal test, at very least we would need to obtain representative samples of black and white students prior to determining whether they chose to attend a historically white or black university.

That aside, Variable C is also shown to affect Variable D but *indirectly* through a mediating variable, namely Variable B. Additionally, Variable D is believed to be influenced directly by an interactive variable, Variable BC, representing the simple first-order interaction at respective levels of D. Variables B and C are sending dotted arrows to interactive Variable BC to denote the fact that in theory, and in many log-linear models, the first-order interaction is independent and hence not affected by its lower order relatives. Yet, some connection, such as the use of dotted arrows, seems appropriate to remind us that Variables B and C need to be crossed and present prior to the realization of the interaction in question. Interactive variables are not commonly seen in either conventional or log-linear causal diagrams, but since this particular interaction happens to reflect McLean's principal research hypothesis—that black students would show a higher rate of on-time graduation at the black university—and since a modicum of complication is forgivable in a last chapter, we accept it as a challenge and hence incorporate it into our model.

Having diagrammed the linkages between and among important variables, two methodological steps remain: (a) verification or rejection of each effect in the model and (b) an overall assessment of the adequacy of the model, or a revised version of the model, to fit observed data. Before implementing these steps, notice that an important procedure in conventional path analysis was not mentioned. Specifically, the systematic decomposition of correlations into direct, indirect, and spurious effects—as described by Asher (1976, pp. 35–44) or Kerlinger and Pedhazur (1973, pp. 314–317), among others—cannot be done in the present circumstances because qualitative variables generally do not lend themselves to product-moment correlations. Nevertheless, there remain a number of working parallelisms to render the log-linear version of path analysis of value for testing theory.

The path diagram in Figure 7.1 presents a number of hypothesized causal links or paths that should be tested in addition to a number of interesting restrictions (relations or effects which have been hypothesized to be nil) that should be examined also. Substantively important paths that have been set to zero are just as important to the integrity of the theory as are the effects explicitly posited in the model. In fact, models become increasingly of interest scientifically as they become more restricted, as they omit paths, because the more restrictions, the greater the number of opportunities to reject statistically, and hence the more valuable the model becomes to science in the long run.

Establishing Paths and Path Coefficients

Putting philosophy aside, it may be recalled from study elsewhere that the basic approach taken in a recursive path analysis is to formulate a predictive equation (or a series of equations in the log-linear) for each endogenous variable. The equation or equations that predict the endogenous variable contain all antecedent variables that are said to be having a direct effect on the predicted variable. Solutions are found for each equation in a series of stages moving from left to right in the recursive system. At each stage, coefficients that measure *partial associations* between antecedent variables and the endogenous response variable are calculated and tested for either statistical or practical significance, or both. For each partial association that is deemed significant, a path from the antecendent variable to the response variable is established and the coefficient is placed on the path to indicate the strength and direction of the causal inference. A similar approach is followed in log-linear path analysis.

The Two-dimensional Stage. With respect to Figure 7.1, we proceed in the standard manner by proceeding from left to right. On the left we find exogenous Variables A and C with no explicit linkage between them. This is appropriate under two sets of circumstances. First, a representative sample in which only n was arbitrarily fixed by the investigator was obtained, and the investigator's theory held that Variables A and C were *not* associated. Or, second, sampling was conducted in such a way that the frequencies in $[AC]$ were fixed or determined by the investigator. The latter would rule out any attempt to assess a relationship between these variables, or in a logit-model analysis to determine whether one variable affects another. McLean fixed the f_{ik}^{ac}'s in the two-way table, which precludes a serious analysis of Variables A and C. However, as an exercise, suppose that McLean had obtained a representative sample of college students and that subsequently he had jointly classified them by ability and race. This would support a general log-linear analysis on the fourfold table. Thus, using the methodology described in Chapter 4, the model for independence and the saturated model could be fitted to $[AC]$. If the latter were accepted, the model would be revised by connecting A and C with a curved double-headed arrow. To indicate the strength of the association, since we are dealing with a fourfold table, Yule's Q defined by Eq. (4.26) would be an appropriate indicator.

But suppose instead that McLean arbitrarily fixed only the levels of one of the variables, say Variable A. Here, Variable A would be considered to be exogenous and explanatory while Variable C would be endogenous and a logit response variable. The null logit model (i.e., the independence model in the symmetrical case) and the model representing main effects for Variable A would be fitted and in turn evaluated. If the main effects model were to be accepted, a straight arrow would be drawn from explanatory A to response C. To suggest the strength and direction of effects, the numerical value of λ_{11}^{ac} from the saturated model is placed on the arrow and serves a function analogous to that of a path coefficient in an analysis done by regression. In addition, it is suggested that the probability value (p-value) associated with the z test on λ_{11}^{ac} also be placed on the path.

The results of an analysis performed on $[AC]$ are summarized in the top portion of Table 7.6. The L^2 from the fit of the independence model was $L_1^2(1) = 86.04$, $p < .00$. This would cause us to adopt the saturated model. But again, this analysis was performed simply as an exercise, for McLean's sampling procedures do not support either a general or a logit-model analysis.

Three-dimensional Stage. Moving to the right, Variable B is the first endogenous variable to be encountered. Assume that McLean did not fix the levels of this variable during sampling, and recall that B_1 contained students who attended the historically white university while B_2 contained students who attended the historically black university. Now, the object is to write and solve equations in which Variable B is the response variable and Variables A and C are predictor variables. That is, the idea is to fit a series of logit models for endogenous Variable B so that we can determine whether: (a) Variable A is having a direct effect on Variable B, subsequent to partialling from A effects shared with Variable C; (b) Variable C is exerting a direct effect on B holding Variable A constant; (c) independent of the above, interaction exists between A and C that in turn functions as a variable affecting B. Even though our model does

TABLE 7.6. Component L^2's for Models Fitted to McLean's Data in the Performance of a Causal Analysis

No.	Marginals Fitted	Residual Comparison	Component L^2	df	p
Preliminary Stage: The fit of models to the $[AC]$ table					
(1)	A, B. (independence)				
(2)	AB. (saturated)	(2)-(1)	86.04	1	.00
Stage 1: The fit of logit models to the $[A\bar{B}C]$ configuration					
(3)	$AC, \bar{B}C.$				
(4)	$AC, \bar{B}C, A\bar{B}.$	(3)-(4)	30.83	1	.00
(5)	$AC, A\bar{B}.$				
(6)	$AC, A\bar{B}, \bar{B}C.$	(5)-(6)	591.05	1	.00
(7)	$A\bar{B}C.$	(6)-(7)	1.27	1	.26
Stage 2: The fit of logit models to the $[ABC\bar{D}]$ table					
(8)	$ABC, B\bar{D}, C\bar{D}.$				
(9)	$ABC, B\bar{D}, C\bar{D}, A\bar{D}.$	(8)-(9)	9.37	1	.00
(10)	$ABC, A\bar{D}, C\bar{D}.$				
(11)	$ABC, A\bar{D}, C\bar{D}, B\bar{D}.$	(10)-(11)	.12	1	.72
(12)	$ABC, A\bar{D}, B\bar{D}.$				
(13)	$ABC, A\bar{D}, B\bar{D}, C\bar{D}.$	(12)-(13)	2.40	1	.12
(14)	$ABC, AB\bar{D}, AC\bar{D}.$	(13)-(14)	3.37	2	.17
(15)	$ABC, AB\bar{D}, AC\bar{D}, BC\bar{D}.$	(14)-(15)	2.90	1	.09
(16)	$ABC\bar{D}.$	(15)-(16)	2.53	1	.11

not hypothesize that an interactive AC variable precipitates change in B, sound analysis would have us search for its possible presence.

Models fitted to the three-way $[ABC]$ table that are needed to implement our work at stage 2 are shown in Table 7.6. Decision making begins with Model (7) where we examine its component chi-square, which is the difference in residuals between the seven-parameter full first-order model ($L_4^2 = L_6^2 = 1.27$) and the eight-parameter saturated model ($L_7^2 = 0.00$). These models are being compared to determine if λ^{abc} is needed to achieve reasonable fit. If it is, then our model will be revised to acommodate this interactive variable. But since the component does not approach significance

$(p < .26)$, we accept the absence of an interactive AC variable affecting Variable B.

The tests of direct effects from Variables A and C are tests of partial, not marginal, associations. As a case in point, to determine whether there is evidence for a direct path from C to endogenous B, we determine whether there is an association between these two variables after partialling from this association the contribution, if any, of Variable A. From our work in Chapter 5 we know that to test for a partial association, λ^{bc} is entered as the seventh or last term in the full first-order model and the fit of this full model is compared to the fit given by a "comparable" six-parameter model, comparable in all respects except for the presence of λ^{bc}. The comparison under discussion is made when the residual for Model (6) in the table, $L_6^2(1) = 1.27$, is subtracted from the residual fit of Model (5), $L_5^2(2) = 592.32$. The component that results, and that is used to test the partial association between C and B, is obviously large, $L_{5-6}^2(1) = 591.05$. A direct connection between C and B has been established.

A suitable coefficient for the path would be the value of λ_{11}^{bc} taken from the full first-order model. Its value is $\lambda_{11}^{bc} = -1.12$. Also on the same path, consider putting the ratio of lambda to its standard error, found to be $z = -18.44$, for the effect in question. If McLean's sampling permitted such an analysis, at this point we would conclude that a student's race affects his or her choice of university. To be more precise, black students (C_1) tend to go to historically black universities (B_2) and vice versa.

Let us see if we can also substantiate a path from Variable A to response B. The strength of the partial association between these variables is reflected in the size of the component resulting from a comparison of Models (3) and (4). It turned out that $L_3^2(2) = 32.10$ and $L_4^2(1) = 1.27$. Thus, $L_{4-3}^2(1) = 30.83$, $p < .00$. If sampling were proper, a path between A and B could be established. Ability level would be said to influence choice of university. Here, λ_{11}^{ab} from full Model (4), or equivalent Model (6), would serve as the path coefficient and concurrent presentation of its associated z statistic would enhance our appreciation of the effect.

A few comments about path coefficients are in order. To start with, we concede that the synonymity between lambdas as path coefficients and partial regression coefficients (betas) in least-squares is less than perfect. To a degree, however, lambda path coefficients are amenable to similar interpretation. Goodman (1972a, 1979), for example, interprets lambda path coefficients as partials in terms of conditional odds. Consider in the example the effect of Race (Variable C) on Type of University (Variable B). Since the variables are both dichotomies, and since we have ruled out an interaction between A and C operating on B, by multiplying the lambda path coefficient by 2—by converting a binomial logit to a full logit—we can obtain the log of the odds of being black and attending the predominantly white university, holding ability level constant. For the example, the logged odds are $2(-1.12) = -2.24$. Obtaining the antilog of -2.24, we can say that the odds of being black and being enrolled in the white university are about .11 to 1, holding levels of A constant. Or, from the opposite perspective, since $2(\lambda_{12}^{bc}) = 2.24$, the odds of being white and being found in the predominantly white university are about 9.39 to 1. To the extent to which Variables A and C interact, however, the interpretation of effects in terms of the conditional can be misleading. Also realize that this manner of interpretation is appropriate only when both the antecedent and response variables are dichotomous.

The direct interpretation of lambdas as logged ANOVA-like effects, however, remains our preference. To many, especially those trained to conduct experiments, the direct interpretation of lambdas presents few if any conceptual difficulties. At very least, the magnitudes of lambda path coefficients can be used as relative indicators, within a given stage, to compare the strength of effects leading to the endogenous variable. Moreover, the direct approach can be extended more easily to paths between polytomous variables.

But irrespective of interpretation, when a path is drawn between variables that are not exclusively dichotomous, because multiple lambda coefficients arise, so do complications. This is yet another problem to be added to our growing list of log-linear applications that need further study. Until more satisfactory path measures become commonly known, it is suggested that the numerical values of component chi-squares be placed on the paths and, accompanying this analogue, the p-values of corresponding component chi-squares. Together, they at very least provide good relative measures of the strength of paths that lead to the same endogenous variable.

The Four-dimensional Stage. Finally, the ultimate response variable, Variable D, is encountered. Main Variables A, C, and B have direct paths leading to D. Also, it has been hypothesized that there is a path between interactive Variable BC, a variable that comes into being because of main Variables B and C but is considered to be independent of them. To assess these many effects on D, logit models in which

D is the logit variable need to be constructed, fitted to the observed [$ABCD$], and subsequently evaluated. The models needed to do the work of this stage are presented, along with their components, at the bottom of Table 7.6.

We look at Table 7.6 first for evidence of an effect on D from the interaction of A, B, and C—a second-order-logit interaction that was, in effect, restricted to zero in the model. When the 16-parameter saturated model was compared to the full second-order model containing 15 parameters—Model (16) vs. Model (15)—the resultant component was found to be $L^2_{15-16}(1) = 2.53$, $p < .11$. Evidence does not appear to be sufficient to contradict the restriction that $\lambda^{abcd} = 0$.

To assess the credibility of the most interesting path in Figure 7.1, that from BC to D, we compare the 15-parameter model, with λ^{bcd}, to the 14-parameter model without λ^{bcd}. The component for the former was $L^2_{14-15}(1) = 2.90$, $p < .09$. Because from the start we regarded this effect to be of greatest substantive interest, and since it does approach traditional criteria for statistical significance, we argue strongly for its continued presence in the model.

The remaining interactive variables that were not explicitly cited in the model are examined collectively through comparison of Model (14), which contains both λ^{abd} and λ^{acd}, against Model (13), which contains all terms in Model (14) with the exception of these two terms. Although not the most rigorous test of the combined influence of these terms, $L^2_{13-14}(2) = 3.37$, $p < .17$. The component is not large enough to reject the restriction that interactive Variables AB and AC have little or no effect on D.

Main variable paths are each tested by model comparisons presented in pairs in the table. As a final example, consider whether there is sufficient support for rejecting the null hypothesis that Variable C has no effect on D, holding Variables A and B constant. As we have done repeatedly, the full first-order-logit model, Model (13), is compared to Model (12), a model with all relevant terms except λ^{cd}. The component observed for Model (13) was not significant ($p < .12$). Unless there are convincing arguments to the contrary, this path will be trimmed from the model when it is revised. Moreover, the revised model will unquestionably not show a path from B to D ($p < .72$), but the path from A to D ($p < .00$) will be retained.

Before the results of our theory trimming are displayed, consider again the coefficients that might appear on the remaining paths. A most conservative approach would be to use lambdas, where appropriate, found from the fitting of the saturated model. But, if we have confidence in our efforts to trim the model, lambdas from

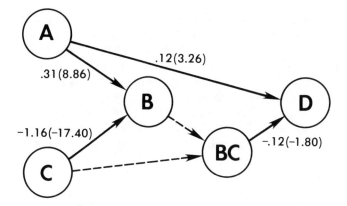

Figure 7.2. Revised path diagram for the McLean study with path coefficients (lambda effect parameters) and statistical tests (z statistics), in parentheses, performed on the coefficients.

the most acceptable logit model can be used. Using the definition for acceptability exercised so often in this text, we choose to take respective Stage 2 coefficients from a model obtained by fitting observed marginals [ABD], [BCD], and [AD]. Our revised path diagram is seen in Figure 7.2.

Testing Models for Goodness-of-Fit

Before a serious investment is made in a path model, the model's ability to reproduce observed cell frequencies should be evaluated. Admittedly, specific judgements about the significance of specific links have been made at every stage, but restrictions that were also accepted at each stage can, when taken together, reduce greatly the model's ability to generate expected frequencies that reasonably approximate those that have been observed. And, obviously, substantive investments in poor-fitting models are to be discouraged. So subsequent to model trimming, we try to judge how well the path model fits observed data by constructing a log-linear model and assessing that model for overall goodness-of-fit.

Assume that we are reasonably satisfied with the path model displayed in Figure 7.2. As mentioned, the task then becomes that of evaluating the congruence between it and observed data. To do this, we will build a log-linear model that tries to incorporate all sampling features, relations, and effects that are part of the path model. The path and log-linear models are not coincidental, for in the log-linear it will not be possible to capture the temporal sequencing and network intricacies of the path model. Sometimes, the log-linear model may of necessity contain surplus effects. Suppose, for example, that higher-order effects like a BC interaction are posited in the path model, and that a conventional computing algorithm, such as the Deming-Stephan, is used. Because the higher order interaction is built into the log-linear model, its lower order relatives, which may not have been cited in the path model, will nevertheless appear in the log-linear representation.

To begin building the model, one first determines what has been fixed as part of the sampling procedure. Differences in a main marginal or among tabular cells that are due to sampling decisions must be taken into account; they must be built into the log-linear model. Let us continue to operate on the assumption, albeit a false assumption, that McLean fixed $[AC]$ during sampling. As a consequence, the model under construction will fit $[AC]$ and contain λ^{ac}.

Next, move to the work at the first stage, ignore the fact that logit models were fitted here, and pick up all relations and effects that have been judged to be significant. In our work, paths were established between Variables A and B and between C and B. Hence, the emerging model will accommodate the fitting of $[AB]$ and $[BC]$. Finally, move to the last stage. Again, ignoring the fact that logit models were used to detect effects, we simply identify established paths. In our example, it is evident that the model should fit $[BCD]$ and $[AD]$.

Putting everything together, the model that represents the path diagram shown in Figure 7.2 will be one that is produced by the fitting of observed marginals $[AB]$, $[AC]$, $[AD]$, and $[BCD]$. When the F_{ijkm}'s generated by this 12-parameter model were compared to f_{ijkm}'s, the residual chi-square was $L^2(4) = 6.60$, $p < .16$. Further, an examination of standard residuals revealed that they were not pronounced; they ranged from -1.1 to $.9$. The fit is indeed satisfactory.

Incidentally, if it can be said that there is a "best" or ideal fit here, it would be one in which the residual chi-square was equal to 4, the number of degrees of freedom, since from Chapter 3 we know that $E(\chi^2) = \nu$. There is a prevailing belief that as a computed residual chi-square becomes less than ν in magnitude, an increasingly better fit is indicated. This is a spurious belief. What one must remember is that if the model were indeed "true," in the long run 50% of computed statistics would be expected to exceed ν and 50% would be less than ν. Therefore, models fit less well as the computed chi-square deviates from ν *in any direction*.

In any event, just because McLean's data happened to be consistent with our log-linear model, this fact alone does not mean that we have substantiated the network of relations and effects in the path model. After all, the same set of variable relations could be rearranged so as to depict a different causal scenario, yet the same log-linear model, hence the same fit, would result. Finally, if we had been serious in our efforts to trim and test the model of this section, the outcome would have been encouraging. If we had not been engaged in a heuristic exercise, we would possess a model with ample opportunities for rejection and refinement in future work. And, during the interim, we would possess a tentative set of relations to help us understand better the nature of the problem.

ADDITIONAL CONCERNS

During the last decade, log-linear methodology has done much to ameliorate a number of problems that formerly plagued the analysis of cross-tabular data. From our reading of this chapter, however, we realize that a number of problems remain. There are two in particular that are stubborn and important and thus merit special recognition. They are the occurrence of "too many" empty cells, and the heavy reliance that we all seem to place on tests of statistical significance during decision making. In this section, brief discussions are given of these and other problems, discussions designed more for edification than for the attainment of solutions.

Small and Zero Cell Frequencies

Problems with zero cell frequencies should not be confused with those associated with very small samples that give rise to *small cell frequencies*. Concern for the latter is historic and centers about the ability of the chi-square distribution to approximate satisfactorily the multinomial (or product multinomial) when *expected* cell frequencies are small. What is small? Unfortunately, authorities differ on the expected cell size that can be used comfortably although the so-called rule of five, mentioned in earlier chapters, can be taken as a reasonable average. Authorities also differ on the merits of the well-known correction offered by Yates (1934) to improve the correspondence between the smooth continuous chi-square and the step-like discontinuous multinomial when samples are small. Some maintain that Yates' correction for continuity should be used routinely in small sample situations (Everitt, 1977), others state that the resulting chi-square statistics are too conservative (Camilli & Hopkins, 1978; Grizzle, 1967), while one study has shown that Yates' correction is too conservative down to a certain sample size (the size depends on population proportions that were permitted to vary in this study), but as sample sizes became smaller the correction resulted in an excess of Type I errors (Wenig, 1979). Most agree, however, that small expected cell sizes promote the same concerns for log-linear analysis as they do for the more traditional forms of qualitative analysis, and that problems are mitigated as the number of cells in the table increase.

But remember, small expected cell frequencies adversely affect only one component of a typical log-linear analysis—the accuracy of the residual and component chi-square statistics associated with the various competing models. Since as models become more restrictive, on average fewer small expected cell frequencies will be generated, so it would seem reasonable to assume, at least for the time being, that the chi-squares of parsimonious models will be less biased than those given by fuller models. In conclusion, although a legitimate concern, there is more to decision making in a log-linear analysis than acceptance or rejection of a hypothesis based on a single test statistic; therefore, in the larger arena of log-linear analysis, the problematic effects of small cell frequencies are not of paramount concern.

Contingency tables that contain void cells can be a major concern, however. Discussion of the problems cannot begin without first making the distinction between *structural* zeros (sometimes called fixed or a priori zeros) and *sampling* zeros (sometimes called random zeros). A structural zero appears in the cell of a table because in the population no such subjects exist. If a table contained a cell for pregnant males, excluding the possibility of clerical error, at the moment at least, that cross-classification would contain a zero. Sampling zeros, on the other hand are not inherent; one will appear only because the overall sample was not large enough to have drawn into it a representative of a particular cross-classification, a cross-classification that likely contains a small relative proportion of the population. Thus, sampling zeros decrease in occurrence as overall sample size increases. Recall that we saw a sampling zero in the data reported by Holton and Nott (Table 5.7).

The distinction is important because the two types of zeros are not equal, at least not equal in their implications for analysis. Sampling zeros, unless numerous, constitute more of any annoyance than a defect. Structural zeros, in contrast, beget tables that are *incomplete* and that require analysis by methods beyond the pale of this introductory book.

Sampling Zeros. Fienberg (1977, p. 108) has pointed out that one of the most powerful properties of log-linear models is that in the presence of sampling zeros, provided that they are not excessive in number, cells with zero observed entries can be given nonzero maximum-likelihood expectancies. Thus, putting aside complications such as too many zeros or zeros that are inherently structural yet treated as sampling, the operational consequences are largely arithmetic and easily handled. As a case in point, consider the calculation of likelihood-ratio chi-squares, using Eq. (3.11), and the fact that an observed frequency of zero can be divided by a nonzero expected frequency with the result being a zero; however, the log of zero is $-\infty$, which is unmanageable in practice. To overcome this and other related problems, rather widespread acceptance has been given to Goodman's suggestion that a small innocuous quantity such as 0.5 be added to all elementary cell frequencies in the observed table. On both the ECTA and BMDP/4F computer programs, for example, by requesting the option DELTA, 0.5 will be added to observed cell frequencies. Relatedly, Goodman (1970) has even suggested that, when fitting a saturated model, 0.5 be added routinely to observed cell frequencies, for this minor correction tends to reduce both the asymptotic bias and standard errors of lambda effect parameters in this particular model. In sum, the addition of a "delta quantity" does much to overcome difficulties with the chi-square. And since with our iterative fitting algorithm expected cell frequencies are obtained first, problems with the estimation of effect parameters will only occur if sampling zeros are so numerous that one or more expected cell frequencies converge to zero.

When sampling zeros are numerous, one is apt to confront problems of (a) not having sufficient information upon which to support an inference, (b) not having sufficient statistical power to drive tests of significance, and (c) not having sufficiently satisfied the assumptions that underlie the proper use of the chi-square which places great strain on the ability of the chi-square to approximate the multinomial. If that were not enough, zeros can be so distributed as to cause, in theory, expected cell frequencies to be negative. In practice, iterative fitting will result in values that converge to zero for these anomalous cells. Finally, in smaller tables particularly, the presence of zeros in common levels of a table can result in a zero entry for the margin of table. When this occurs, modifications need to be made in both the fitting algorithm and the rule for determining the proper number of degrees of freedom. These latter modifications are discussed by Fienberg (1977, Chapter 8).

Structural Zeros.   As mentioned, zeros are considered to be structural if the cross-classification does not exist (e.g., normal deviates) or if we intentionally force cells to have a zero frequency during iterative fitting and residual testing as we did when working with models for symmetrical change. It is the former type of zero entry that causes difficulties because they make contingency tables incomplete. Incomplete tables are not encountered frequently in the literature; when they are, they are more likely to be seen in medicine and biology than in the social sciences. Undoubtedly, this difference reflects, in part, the general nature of the subject matter and possibly the greater ease with which social scientists can combine or eliminate classes to avoid missing cells. In any event, as the reader has surely surmised, incomplete tables are to be avoided if possible but not at the expense of treating a structural zero as if it were a sampling zero or altering seriously the nature of the problem by collapsing over categories until zeros vanish.

One course of action that might work for a large incomplete table is to break it up into smaller configurations that make contextual sense and then to subject the smaller tables to separate but normal log-linear analyses. If this cannot be done meaningfully, the table must be considered incomplete and therefore only amenable to analysis by *quasi-log-linear models*. Quasi-log-linear models are analogous to the models that we have been using on complete tables. For example, it is possible to structure and fit a model for *quasi-independence*, a model that generates expected cell frequencies for independence in the usual way except that columns (rows) containing one or more zeros are not considered. Understandably, with more constraints imposed upon the data, it takes longer to determine whether the table can be analyzed and with less assurance that it can be. Recent, brief, and relatively understandable discussions of incomplete tables and quasi-log-linear models are offered by Fienberg (1977) and Upton (1978), though an early paper by Fienberg (1972) still merits the highest recommendation.

Chi-square Values and Sample Size

Perhaps the greatest limitation of the approach that we have taken to log-linear analysis has been our heavy reliance on the statistical significance of chi-square values when faced with decisions about model appropriateness. At issue, of course, is the fact that the magnitude of a chi-square statistic and thus its p-value is a function of sample size. If a sample can be made large enough, any model, except the saturated model, can be rejected on the basis of a statistical criterion because ultimately an n will be attained whereby the residual chi-square will be significant. On the other hand, the probability of accepting a model increases as n decreases. Fortunately, subject matter researchers are becoming increasingly aware of the relationships between statistical significance and sample size and are formally or informally adjusting their inferences accordingly. In some areas of analysis, statistical power tables and/or measures of association that are independent of n are available. Unfortunately, in our area, as soon as we move beyond two-way tables, there are no widely accepted measures of association to help with the assessment of model fit.

The problem is most pronounced in the symmetrical mode. When inquiry is truly symmetrical, we search for the most restricted model that still represents a plausible explanation of relations in data. Hence, *acceptance* of restrictions (i.e., acceptance of null hypotheses) plays a major role, yet we know that in statistical decision theory, acceptance of a restriction is a weak form of decision. It is so weak and so tentative that many would prefer that we do not even speak of "acceptance" but instead would have us say that we are "unable at the moment to reject." In any event, in earlier chapters we operated in the symmetrical case, and the strategy that we used (see Figure 5.1) had us first examine residual L^2's to identify models that did not yield large and significant L^2's and therefore did not fit the data poorly. These models were retained for further examination.

While implementing the strategy, we were aware that as samples become small, the likelihood will increase that all models, even the null model, will be retained as candidates for further study. Of course, as samples become large the reverse is true; in the extreme, only the saturated model will be retained. Granted, the skillful analyst will not rely exclusively on residual chi-squares during this initial step. He or she will be constantly monitoring the distance between chi-square values and their numbers of degrees of freedom for $E(L^2) = \nu$. In addition, analyzing standardized or Freeman-Tukey deviates can be insightful. But realize that these ancillary aids in the selection process are similarly affected by sample size. Further, decisions made during the second phase of the selection process, the phase in which component chi-squares are tested for significance, are also affected by the size of samples.

Goodman (1971, 1972b) has proposed a relatively straight-forward coefficient that merits consideration by researchers who frequently find themselves working in the symmetrical mode. It is analogous to a semi-partial correlation in the framework of a forward-solution multiple regression, it is independent of sample size, and some (Zahn & Fein, 1979) have found it to be of value in their attempts to judge the goodness-of-fit of general models. The idea is to identify a baseline model, determine the residuum about that model, and then determine how much of that residuum can be explained by more saturated models of interest. It is most common to choose the model of mutual independence, say Model (0), as the baseline model, although this need not always be so. Then, for a less restricted model of interest, say Model (i), Goodman's coefficient for the ith model is

$$\Lambda_i = (L_0^2 - L_i^2)/L_0^2 \tag{7.7}$$

The coefficient denoted by uppercase Greek lambda ranges from 0.0 to 1.0 and for Model (i) can be interpreted as the proportion of total variation in the residuum about the baseline model that can be explained by all nonbaseline terms in Model (i). Thus, irrespective of statistical significance, as models become less restrictive and begin to account for between 80 and 90% of meaningfully defined residual variation, goodness-of-fit is suggested. Unfortunately, minimum specific criteria for selecting models by this means have yet to be formulated.

When inquiry is symmetrical, therefore, we cannot escape the fact that ultimately the notion of acceptance of a null enters the picture. It is so critical that substantive considerations must be present, operative, and prominent in the model selection process. But when inquiry is asymmetrical, our strategy for decision making strikes a different emphasis, that of *rejection*—specifically, the rejection of component chi-squares if they are justifiably large. Unlike the statistical decision to accept, rejection is a strong, albeit conditional, decision. The size of samples affects the size of components and hence their p values, but in a less beguiling way. Small samples promote Type II errors; large samples promote the ascription of statistical significance to effects that may not be practically significant. Nevertheless, we find ourselves on more familiar and comfortable ground because rejection of a null is the requisite for advancing a finding, and the burden for rejection, as statistical tradition would have it, is on the shoulders of the researcher.

THE NEED FOR LOG–LINEAR METHODS

If a table's dimensionality exceeds two (i.e., $k > 2$) and inquiry is symmetrical, generally it is advantageous to analyze resultant data with log-linear methodology. The older methods of symmetrical analysis surveyed in Chapter 4 lack the comprehensiveness of log-linear analysis; they do not accommodate the simultaneous analysis of more than two variables and therefore are not able to address important concepts such as conditional independence or partial association. Further, at this writing, there does not appear to be a contemporary or emerging system of analysis for qualitative data with advantages superior to those of log-linear when $k > 2$ and the mode is symmetrical.

The eminence of log-linear analysis has been questioned, however, when the mode of inquiry is asymmetrical. Many have asked: Why bother with log-linear theory and technique when suitable alternative techniques appear to exist, techniques that are well established in the service of asymmetrical inquiry? In short, do we really need log-linear analysis?

Before we respond to the central question, techniques that might appear to be serious competitors to logit-model analysis merit, at least, acknowledgment. While reading accounts of the studies conducted by McLean and Peters, it may have crossed the reader's mind that instead of a logit-model analysis, McLean conceivably could have used multiple regression (MR) as his principal analytical technique, and Peters could have used canonical variate analysis (CVA). Both are established parametric procedures and are described well in a number of tests on multivariate analyses (e.g., Green, 1978; Harris, 1975). MR, in particular, is well known among behavioral scientists, appearing frequently in the research literature. Although not as frequently seen, CVA is to be appreciated for its catholicity because the majority of quotidian parametric techniques—for example, t tests, PPM r's, ANOVAs, MANOVAs, and MR's—can be shown to be special or reduced cases of this more generic procedure.

At any rate, and at first glance, MR would appear to be a suitable and attractive alternative to log-linear when the response variable is dichotomous, as it was in the study by McLean. On the other hand, when response is polytomous, as it was in Peters' study, CVA would appear to be an attractive alternative, or so it would seem.

We consider first the polytomous response and hence the potential use of CVA. Without going into great detail, a CVA can be performed when two different *sets* of measured variables are available on a group of subjects. Let the variables in one of the sets be denoted by X, and be indexed by j, such that $j = 1, 2, \ldots, p$. Variables measured on the group of subjects that comprise the second set will be denoted by Y where subscript j varies from $1, 2, \ldots, q$. In theory, the object of CVA is to try to account for the relations *between* the X_j's and Y_j's by solving for a small number of underlying linear combinations in pairs (pairs of canonical factors) and then, if possible, to try to interpret substantively the canonical factors that are "explaining" the between-set relations. Most often, in both exposition and practice, all variables in both sets are considered to be interval or ratio in nature, yet they need not be.

Some or all of the X_j's (or Y_j's) in a CVA can be qualitative. In the Peters' study, for example, the four-level variable of response could have been subjected to one of several existing *coding methods* (e.g., dummy coding, effect coding, orthogonal coding, etc.)—as is frequently done to qualitative independent variables when MR is used to perform an ANOVA—and the resultant three coded variables could be placed in the set of Y_j's, the dependent variable set. The four explanatory variables in Peters' study were also amenable to coding by similar means; thus they could have been incorporated into a set of X_j's, a set of coded independent variables. (If interactions are included, p would equal 7 in this set.) In sum, structurally it is possible to satisfy the layout of a CVA even if both sets of variables are coded qualitative variables, and mathematically it is possible to perform the analysis so as to produce canonical correlations, tests on canonical correlations, structure loadings of X_j's and Y_j's on respective canonical factors, and measures of redundancy, among other statistics. In the hands of a skilled data analyst, the statistical production of a CVA can yield insights into the number and nature of relationships between explanatory variables and the response variable. The substantive interpretation of a CVA, however, demands a relatively high level of knowledge and skill.

Less complicated is MR, an alternative to a logit-model analysis when the response variable is dichotomous. Recall that McLean's response variable was a dichotomy. He therefore could have assigned the number 1 to all students who graduated "on time" and 0 to those who did not. By doing so, the original dichotomy would have been transformed to a univariate response that could serve as the dependent variable in a regression, a regression in which the independent variables would be coded explanatory variables. Actually, the analysis being described can also be thought of as a three-way ANOVA on a dichotomous dependent variable—or a CVA in which only one coded variable makes up the Y set (i.e., $q = 1$). In any case, least-square, not maximum likelihood, is the basis for estimation, and proportions, not logits, are the dependent variable. Given that least-squares regression is at present highly developed and its computations in our computer age are most manageable, is there really anything to be gained by employing the methods of this book?

The answer to the question above is *yes*. Some of the reasons in support of the simple answer are in turn simple, others are not; some are rooted in statistical theory, others are rooted in application; and one argument even approaches the philosophical. We will entertain them in the order suggested.

Arguments Based on Theory

Since CVA and its special case, MR, are parametric procedures, by definition certain assumptions are made about the distribution of variables in the population prior to mathematical development. Moreover, in practice, violations of these assumptions can lead to biased tests of statistical significance. Recall that *normality* and *homogeneity of variance* are two of the assumptions that are basic to parametric testing. When response is truly qualitative, analysis of this response with either CVA or MR will likely contradict both of these basic assumptions.

For expository simplicity, consider a dichotomous response variable such as "pass" (coded as a 1) and "fail" (coded as a 0), and an explanatory variable such as sex that also can be dummy coded readily. Mentally place these coded variables into a regression equation, or perceive the situation to be one where a t test will be performed on the dichotomous pass-fail variable. Both approaches are equivalent and both assume that the dependent variable is distributed normally in the population of females and males, and that the variances of the dependent variable for females and males are equal in these populations. But remember, in a logit-model analysis we want to know whether the profiles of *proportional* response are similar or different by gender. Therefore, in the current context the basic assumptions are (a) that the proportion of 1's (passes) and 0's (failures) are normally distributed in each sex group and (b) that the variance of proportional response to the pass-fail variable is equivalent over sex groups. Now the first assumption can be tenable only when the distribution of 1's and 0's is about even, say a "50-50 split," within both the group of females and the group of males. The second assumption concerns within-group variance, which is a direct function of the distribution. If P stands for the proportion of 1's within a sex group, then the group's variance is equal to the product $P(1 - P)$. Therefore, only when there is little or no difference in the response of males and females will the assumption of equal variance be tenable.

In sum, to the extent to which response to the dichotomy deviates from an even distribution within groups, and to the extent to which response patterns are not similar within groups, both assumptions will be violated. To compound matters, in the presence of violations often it is difficult to determine the magnitude and direction of resultant bias in test statistics. And as one would expect, the problem is magnified with an increase in the number of explanatory variables and when a CVA is performed because response is polytomous. Fortunately, a logit-model approach to the analysis is far less dependent on the distributional assumptions of normality and homogeneity of variance.

There is still another fundamental problem with dummy variable regression. Briefly, in regression least-squares procedures are used to solve for regression weights, which in turn are used in the equation to generate predicted values. The predicted values here are proportions but their range is unrestricted. It is theoretically possible, therefore, for the regression to fit cell proportions that are greater than unity or even proportions that are negative. Of course, there is nothing comparable in the reality of our subject matter, a reality where proportions find limits between zero and unity. Needless to say, unrestricted proportional estimates constitute a disturbing and unnecessary technical liability in the present context.

Arguments Based on Practice

Turning to points that are slightly more relevant to application, in comparison to its competitors a logit-model analysis tends to fit observed data better, can be more powerful, is much easier to interpret when response is polytomous, and is relatively self-contained in that it possesses built-in follow-up devices. For readers of the preceding three chapters, few words need be written about the latter advantage, for repeatedly we have seen how estimates of effect parameters, especially lambda parameters in linear models, can be used to assist in the explication of an omnibus result. Most salutary too is that the lambda effects can be tested for statistical significance and that the size and patterns of z statistics reveal much about the specific nature of results.

Closely tied to follow-up is the matter of substantive interpretability. The fact of the matter is that some statistical techniques are more readily understood, interpreted, and communicated in the language of the subject than are others. In general, it is fair to say that multivariate techniques have earned reputations for being difficult to interpret substantively, although strict mathematical interpretations are not difficult to make. Most prominent among multivariate procedures that are difficult to interpret is CVA. Granted, the significance of a canonical correlation coefficient is an indication that one or more effects is also significant, but remaining statistical outputs (e.g., canonical weights, structure coefficients, etc.) are more suited to the study of complex *relationships*, not effects. In short, a CVA is difficult to interpret in its own right, but even greater difficulties are encountered when an attempt is made to use this vehicle to study group difference.

But before matters of interpretability become a concern, effects must first be found to be significant, either statistically or practically. It may be remembered that in our discussion of ML estimation, it was mentioned that estimates obtained by ML procedures have somewhat smaller variances than comparable least-squares estimates (see Rao, 1965, and Grizzle et al., 1969). Consequently, on average, ML procedures, and thus logit models, will be more efficient; they will give, on average, smaller standard errors about effects. In the past it has been said that the loss in efficiency resulting from the use of least-squares regression was more than offset by the ease and availability of computational routines for regression (Grizzle et al., 1969), but the credibility of this argument will diminish rapidly as log-linear computer software becomes increasingly available.

Another important advantage of logit models is that they tend to fit observed data better than their least-squares counterparts. That is, the differences between observed frequencies and the expected frequencies given by logit models tend to be relatively small. It should be conceded, however, that when observed proportions are all in the mid-range, say between .25 and .75, differences in the numerical values of model parameters, and thus fit, are small. But as observed proportions in a table deviate from mid-range values, the fit and even conclusions can differ between types of models. Magidson (1978) compared several relevant unsaturated least-squares regression models with corresponding logit models within the context of a $2 \times 2 \times 2$ table, and although there were no discrepancies in omnibus conclusions between model offerings, the logit models consistently were shown to fit data substantially better than the regression models. In short, logit models were more accurate representations of observed data. Magidson also illustrated several extreme situations in which the two approaches yielded different omnibus conclusions.

Concluding Remarks

The strongest and best justification for the use of log-linear models in the general case and logit models in the asymmetrical case, however, is that these models are inherently compatible with the data at hand and the concepts that we use to understand those data. In the work described in this book, we have been primarily interested in proportional response. In both symmetrical and asymmetrical situations, ultimately the question has been whether observed elementary cell proportions are similar to expected cell proportions given by a model that represents a hypothesis. And from Chapter 3 we know that proportional response and ML estimation go hand in hand, that ML estimates are both most efficient and sufficient estimators of population proportions. It follows, therefore, that ML estimation—not weighted or unweighted least-squares—is to be preferred when working with proportions in contingency tables.

But there is more. Consider that throughout most of this book we have been dealing with formulations, and thinking with concepts, that are in their very nature multiplicative. Take, for example, a most basic concept such as *independence*. Independence when used within the context of a contingency table is a multiplicative formulation as evidenced by its definition in Eq. (3.6) for two-way tables and Eq. (5.5) for three-way tables. (Even the notion of a significant effect in the asymmetrical is multiplicative, for it represents a significant departure from proportions under the independence hypothesis.) In short, the majority of concepts with which we have been struggling are essentially multiplicative in origin, and we have been attempting to represent these concepts with models that are also essentially multiplicative: log-linear models. Granted, log-linear models are not multiplicative per se. But remember how we began our study of qualitative data. Because concepts were multiplicative, we initially couched our representations in multiplicative "terms" (so to speak). However, we found it easier to compute in the linear, and thus multiplicative factors were transformed to additive terms by taking the logs of the former. The point is this: despite their appearance, log-linear models are fundamentally multiplicative creatures and are therefore true to the tables, data, and ideas that have occupied our attention over these many pages.

COMMENT

Although it is not usual to do so in books like this, this writer feels the need to compliment those who knew little about log-linear analysis at the outset, who have persisted through this text, and who through their persistence have obtained at least a working knowledge of this relatively new and most useful system of analysis. Granted, the journey has been long and at times difficult, but if this text has contributed in part to the acquisition of knowledge sufficient to pursue more advanced expositions of this topic, then both the journey and the preparation of this text have been worthwhile.

APPENDIXES

APPENDIX A

A Proportional Iterative Fitting Algorithm

As log-linear models increase in complexity it becomes more difficult, and sometimes is not even possible, to generate expected elementary cell frequencies directly by using probability theory. In Chapter 5, for example, recall that while fitting models to the adverse impact data, our direct computations of F_{ijk}'s based on principles of probability became more involved as we proceeded from Model (1) to Model (6), as these models were defined by Table 5.2. Then when Model (7) was considered (i.e., the full first-order model), we learned that as a result of mutual dependencies, direct computations of the F_{ijk}'s were not possible. To obtain ML expectancies for this model we had no recourse but to resort to an iterative fitting procedure.

Several iterative fitting procedures are currently available. The two that are most frequently cited and used are the Newton-Raphson algorithm, described in Haberman (1978, pp. 128-130), and the iterative proportional fitting algorithm introduced by Deming and Stephan (1940), adapted and programmed in Fortran by Haberman (1972), and described in a number of sources (Haberman, 1978, pp. 125-128; Fienberg, 1977, pp. 33-36). The former does not require that models be arranged hierarchically, but the latter has advantages of simplicity and wide acceptance. Essentially, it is the latter algorithm that is used in ECTA and BMDP/4F. It was also an adaptation of the Deming-Stephan algorithm that was used in Chapter 5 to generate F_{ijk}'s for Model (7) of the adverse impact example. A brief demonstration of the mechanics of this algorithm is presented in this appendix.

As implied by the adjective *iterative*, the fitting of expected cell frequencies proceeds in a series of successive steps or major iterations. Moreover, each major iteration involves a number of subiterations. Major iterations are executed until reasonable convergence is observed. That is, following each major iteration, resultant expected cell frequencies are compared to prior cell frequencies, and if agreement is close—say within the typical default criterion of .1— fitting terminates. If agreement is not good, another major iteration is performed. The number of subiterations within a major iteration depends on the number of observed marginals that are being fitted by the particular model.

As mentioned, Model (7) has been chosen for the demonstration. It is the model that produced F_{ijk}'s for $[ABC]$ by fitting *observed* frequencies in $[AB]$, $[AC]$, and $[BC]$ as seen in the adverse impact example of Chapter 5. For the demonstration, let m denote an unspecified major iteration where $m = 1, 2, \ldots ,$ convergence; and let s represent a particular subiteration where, in our work, $s = 1, 2,$ or 3. Tailoring a general formula to suit our three-dimensional situation, expected elementary cell frequencies for subiteration s and the mth iteration are given by

$$F_{ijk}^{(m,s)} = F_{ijk}^{(s-1)} \, \frac{f_{xy}^{(s)}}{F_{xy}^{(s-1)}}$$

where $F_{ijk}^{(s-1)}$'s are *expected* cell frequencies from the previous subiteration, $f_{xy}^{(s)}$'s are *observed* frequencies in the particular two-way configuration that is being fitted during subiteration s, and the $F_{xy}^{(s-1)}$'s are *expected* frequencies in the particular two-way configuration being fitted, expectancies obtained from the previous subiteration.

The First Major Iteration. For complete contingency tables, fitting usually begins by inserting the constant 1 (unity) into the elementary cells of the table. This is the initial *preparatory* iteration with the result that each of the eight cells in the three-way table contains a 1. Define this preparatory iteration as $m = 0$.

We are now ready to execute the first major iteration, which begins by collapsing over Variable C to obtain observed frequencies in $[AB]$. Verify that

$$[AB] = (f_{11}^{ab} = 11, f_{12}^{ab} = 9, f_{21}^{ab} = 42, f_{22}^{ab} = 38)$$

Using the formula presented above to compute the F_{ijk}'s for $m = 1$ and the first subiteration ($s = 1$), we find

$$F_{111}^{(1,1)} = (1) [11/(1 + 1)] = 5.5$$
$$F_{112}^{(1,1)} = (1) [11/(1 + 1)] = 5.5$$
$$F_{121}^{(1,1)} = (1) [9/(1 + 1)] = 4.5$$
$$\vdots$$
$$F_{222}^{(1,1)} = (1)(38/2) = 19.00$$

Hence, at the completion of the subiteration, we find that the following expectancies have been produced:

		C_1	C_2
A_1	B_1	5.5	5.5
	B_2	4.5	4.5
A_2	B_1	21.0	21.0
	B_2	19.0	19.0

In the second subiteration, we use and fit observed $[AC]$ which is

$$[AC] = (f_{11}^{ac} = 10, f_{12}^{ac} = 10, f_{21}^{ac} = 30, f_{22}^{ac} = 50)$$

The values of F_{ijk}'s for $m = 1$ and $s = 2$ are

$$F_{111}^{(1,2)} = (5.5) [10/(5.5 + 4.5)] = 5.5$$
$$F_{211}^{(1,2)} = (21.0) [30/(21 + 19)] = 15.75$$
$$\vdots$$
$$F_{222}^{(1,2)} = (19)(50/40) = 23.75$$

In the third subiteration we use $[BC]$. A selected computation in this set is

$$F_{111}^{(1,3)} = (5.5)(25/21.25) = 6.4706$$

At the end of the first major cycle, the cell expectancies are:

		C_1	C_2
A_1	B_1	6.4706	4.8504
	B_2	3.6000	5.0973
A_2	B_1	18.5294	23.1496
	B_2	11.4000	26.9027

At this point, the frequencies immediately above are compared to those that were specified during the initial preparatory fitting, i.e., when $m = 0$. Recall that $F_{ijk}^{(m = 0)} = 1.0$ for all values of i, j, and k when $m = 0$. Since agreement between corresponding elementary cells does not even approach the criterion, a second major iteration needs to be undertaken.

The Second Major Iteration. Again, the first subiteration fits $[AB]$. Using the expectancies from the last subiteration in the first major cycle, as a selective illustration we see

$$F_{111}^{(2,1)} = (6.4706) \ [11/(6.4706 + 4.8504)]$$
$$= 6.2871$$

For $m = 2$, $s = 2$,

$$F_{111}^{(2,2)} = (6.2871)(10/10.0124) = 6.2793$$

For $m = 3$, $s = 3$,

$$F_{111}^{(2,3)} = (6.2793)(25/24.9626) = 6.2887$$

At termination, the expected cell frequencies generated during this major iteration are compared to their counterparts from the first major iteration. We will only compare $F_{111}^{(1,3)} = 6.4706$ with $F_{111}^{(2,3)} = 6.2887$. Neither this comparison nor the remaining seven comparisons were able to demonstrate congruence within one-tenth of a frequency. Therefore, at least a third major iteration will be required.

The Third Major Iteration. Using parallel procedures, three subiterations are undertaken in this cycle. For $m = 3$, $s = 3$, for example, it will be found that

$$F_{111}^{(3,3)} = (6.2882)(25/24.9998) = 6.2882$$

Note that this result compares favorably with that of $F_{111}^{(2,3)}$. In fact, all cell expectancies at the end of the third iteration are within 0.1 of corresponding values emanating from the second iteration. Iterative fitting terminates since the desired F_{ijk}'s for Model (7) have been produced.

APPENDIX B

Critical Values for Chi-square Statistics

	Percentile						
	50	75	90	95	97.5	99	99.9
				p			
df	.50	.25	.10	.05	.025	.01	.001
1	.45	1.32	2.71	3.84	5.02	6.63	10.8
2	1.39	2.77	4.61	5.99	7.38	9.21	13.8
3	2.37	3.11	6.25	7.81	9.35	11.3	16.3
4	3.36	5.39	7.78	9.49	11.1	13.3	18.5
5	4.35	6.63	9.24	11.1	12.8	15.1	20.5
6	5.35	7.84	10.6	12.6	14.4	16.8	22.5
7	6.35	9.04	12.0	14.1	16.0	18.5	24.3
8	7.34	10.2	13.4	15.5	17.5	20.1	26.1
9	8.34	11.4	14.7	16.9	19.0	21.7	27.9
10	9.34	12.5	16.0	18.3	20.5	23.2	29.6
11	10.3	13.7	17.3	19.7	21.9	24.7	31.3
12	11.3	14.8	18.5	21.0	23.3	26.2	32.9
13	12.3	16.0	19.8	22.4	24.7	27.7	34.5
14	13.3	17.1	21.1	23.7	26.1	29.1	36.1
15	14.3	18.2	22.3	25.0	27.5	30.6	37.7
16	15.3	19.4	23.5	26.3	28.8	32.0	39.3
17	16.3	20.5	24.8	27.6	30.2	33.4	40.8
18	17.3	21.6	26.0	28.9	31.5	34.8	42.3
19	18.3	22.7	27.2	30.1	32.9	36.2	43.8

	Percentile						
	50	75	90	95	97.5	99	99.9
				p			
df	.50	.25	.10	.05	.025	.01	.001
20	19.3	23.8	28.4	31.4	34.2	37.6	45.3
21	20.3	24.9	29.6	32.7	35.5	38.9	46.8
22	21.3	26.0	30.8	33.9	36.8	40.3	48.3
23	22.3	27.1	32.0	35.2	38.1	41.6	49.7
24	23.3	28.2	33.2	36.4	39.4	43.0	51.2
25	24.3	29.3	34.4	37.7	40.6	44.3	52.6
26	25.3	30.4	35.6	38.9	41.9	45.6	54.1
27	26.3	31.5	36.7	40.1	43.2	47.0	55.5
28	27.3	32.6	37.9	41.3	44.5	48.3	56.9
29	28.3	33.7	39.1	42.6	45.7	49.6	58.3
30	29.3	34.8	40.3	43.8	47.0	50.9	59.7
40	39.3	45.6	51.8	55.8	59.3	63.7	73.4
50	49.3	56.3	63.2	67.5	71.4	76.2	86.7
60	59.3	67.0	74.4	79.1	83.3	88.4	99.6
100	99.3	109.1	118.5	124.3	129.6	135.8	149.5

REFERENCES

Asher, H. B. *Causal modeling.* Sage University Paper series on Quantitative Applications in the Social Sciences, series no. 07-003. Beverly Hills and London: Sage Publications, 1976.

Bentler, P. M. Multivariate analysis with latent variables: Causal modeling. In M. R. Rosenzweig & L. W. Porter (Eds.), *Annual review of psychology* (Vol. 31). Palo Alto, Calif.: Annual Reviews, 1980.

Birch, M. W. Maximum likelihood in three-way contingency tables. *Journal of the Royal Statistical Society,* 1963, *B25,* 220-233.

Bishop, Y. M. M., Full contingency tables, logits, and split contingency tables. *Biometrics,* 1969, *25,* 383-400.

Bishop, Y. M. M., Fienberg, S. E., & Holland, P. W. *Discrete multivariate analysis: Theory and practice.* Cambridge, Mass.: MIT Press, 1975.

Bock, R. D., & Yates, G. MULTIQUAL: Log-linear analysis of nominal and ordinal data by the method of maximum likelihood. Chicago: National Educational Resources, 1973.

Brown, M. B. Screening effects in multidimensional contingency tables. *Applied Statistics,* 1976. *25,* 37-46.

Camilli, G., & Hopkins, K. D. Applicability of chi square to 2×2 contingency tables with small expected cell frequencies. *Psychological Bulletin,* 1978, *85,* 163-167.

Cochran, W. G. Some methods for strengthening the common χ^2 tests. *Biometrics*, 1954, *10*, 417–451.

Cook, T. D., & Campbell, D. T. *Quasi-experimentation: Design and analysis issues for field settings.* Chicago: Rand McNally, 1979.

Deming, W. E., & Stephan, F. F. On a least-squares adjustment of a sampled frequency table when the expected marginal totals are known. *Annals of Mathematical Statistics*, 1940, *11*, 427–444.

Dixon, W. J., *BMDP statistical software.* Berkeley: University of California Press, 1981.

Dixon, W. J., & Brown, M. B. *BMPD-79: Biomedical computer programs P-series.* Berkeley: University of California Press, 1979.

Everitt, B. S. *The analysis of contingency tables.* New York: Halsted Press, 1977.

Fay, R. E., & Goodman, L. A. *ECTA program: Description for users.* Chicago: University of Chicago, 1975.

Fienberg, S. E. An iterative procedure for estimation in contingency tables. *Annals of Mathematical Statistics*, 1970, *41*, 907–917. (a)

Fienberg, S. E. The analysis of multidimensional contingency tables. *Ecology*, 1970, *51*, 419–433. (b)

Fienberg, S. E. The analysis of incomplete multi-way contingency tables. *Biometrics*, 1972, *28*, 177–202.

Fienberg, S. E. *The analysis of cross-classified categorical data.* Cambridge, Mass.: MIT Press, 1977.

Fisher, R. A. The conditions under which χ^2 measures the discrepancy between observed observation and hypothesis. *Journal of the Royal Statistical Society*, 1924, *87*, 442–450.

Freeman, M. F., & Tukey, J. W. Transformation related to the angular and square root. *Annals of Mathematical Statistics*, 1950, *21*, 607–611.

Glass, G. V., & Stanley, J. C. *Statistical methods in education and psychology.* Englewood Cliffs, N.J.: Prentice-Hall, 1970.

Goldsmid, C. A., Gruber, J. E., & Wilson, E. K. Perceived attributes of superior teachers (PAST): An inquiry into the giving of teacher awards. *American Educational Research Journal*, 1977, *14*, 423–440.

Goodman, L. A. The multivariate analysis of qualitative data: Interactions among multiple classifications. *Journal of the American Statistical Association*, 1970, *65*, 226–256.

Goodman, L. A. The analysis of multidimensional contingency tables: Stepwise procedures and direct estimation methods for building models for multiple classifications. *Technometrics*, 1971, *13*, 33–61. (a)

Goodman, L. A. Partitioning of chi-square, analysis of marginal contingency tables, and estimation of expected frequencies in multidimensional contingency tables. *Journal of the American Statistical Association*, 1971, *66*, 339–344. (b)

Goodman, L. A. A general model for the analysis of surveys. *American Journal of Sociology*, 1972, *77*, 1035–1086. (a)

Goodman, L. A. A modified multiple regression approach to the analysis of dichotomous variables. *American Sociological Review*, 1972, *37*, 28–46. (b)

Goodman, L. A. Guided and unguided methods for the selection of models for a a set of *T* multidimensional contingency tables. *Journal of the American Statistical Association*, 1973, *68*, 165–175.

Goodman, L. A. *Analyzing qualitative/categorical data*. Cambridge, Mass.: Abt Books, 1978.

Goodman, L. A. A brief guide to the causal analysis of data from surveys. *American Journal of Sociology*, 1979, *84*, 1078–1095.

Green, P. E. *Analyzing multivariate data*. Hinsdale, Ill.: Dryden Press, 1978.

Grizzle, J. E. Continuity correction in the χ^2 test for 2 × 2 tables. *American Statistician*, 1967, *21*, 28–32.

Grizzle, J. E., Starmer, C. F., & Koch, G. G. Analysis of categorical data by linear models. *Biometrics*, 1969, *25*, 489–504.

Grizzle, J. E., & Williams, O. D. Long-linear models and tests of independence for contingency tables. *Biometrics*, 1972, *28*, 137–156.

Haberman, S. J. Log-linear fit for contingency tables (Algorithm AS 51). *Applied Statistics*, 1972, *21*, 218–225.

Haberman, S. J. The analysis of residuals in cross-classified tables. *Biometrics*, 1973, *29*, 205–220.

Haberman, S. J. *Analysis of qualitative data: Introductory topics* (Vol. 1). New York: Academic Press, 1978.

Harris, R. J. *A primer of multivariate statistics.* New York: Academic Press, 1975.

Hays, W. L. *Statistics for the social sciences* (2nd ed.). New York: Holt, Rinehart and Winston, 1973.

Heise, D. R. *Causal analysis.* New York: John Wiley, 1975.

Holton, J. & Nott, D. L. *The experimental effects of reflective teaching upon preservice teachers ability to think and talk critically about teaching.* Paper presented at the meeting of the American Educational Research Association, Boston, April 1980.

Jöreskog, K. G. Structural analysis of covariance and correlational matricies. *Psychometrika,* 1978, *43,* 443-477.

Kennedy, J. J. *An introduction to the design and analysis of experiments in education and psychology.* Washington, D.C.: University Press of America, 1978.

Kerlinger, F. N., & Pedhazur, E. J. *Multiple regression in behavioral research.* New York: Holt, Rinehart and Winston, 1973.

Knoke, D., & Burke, P. J. *Log-linear models.* Sage University Paper series on Quantitative Applications in the Social Sciences, series no. 07-020. Beverly Hills and London: Sage Publications, 1980.

Lancaster, H. O. Complex contingency tables treated by the partition of chi-square. *Journal of the Royal Statistical Society,* 1951, *B13,* 242-249.

Lee, S. K. On the asymptotic variances of $\hat{\mu}$ terms in loglinear models of multi-dimensional contingency tables. *Journal of the American Statistical Association,* 1977, *72,* 412-419.

Lohnes, P. R., & Cooley, W. W. *Introduction to statistical procedures: With computer exercises.* New York: John Wiley, 1968.

Magidson, J. An illustrative comparison of Goodman's approach to logit analysis with dummy variable regression analysis. In L. A. Goodman, *Analyzing qualitative/categorical data.* Cambridge, Mass.: Abt Books, 1978.

Marascuilo, L. A., & McSweeney, M. *Nonparametric and distribution-free methods for the social sciences.* Monterey, Calif.: Brooks/Cole, 1977.

Marks, E. Methods for analyzing multidimensional contingency tables. *Research in Higher Education*, 1975, *3*, 217-231.

Maxwell, A. E. *Analyzing qualitative data*. London: Methuen, 1961.

McLean, J. A. *Graduation and nongraduation rates of black and white freshman entering two state universities in Virginia.* Unpublished doctoral dissertation, The Ohio State University, 1980.

Mitroff, I. I., & Kilmann, R. H. *Methodological approaches to social sciences.* San Francisco: Jossey-Bass, 1978.

Myers, I. *Manual for the Myers-Briggs Type Indicator.* Princeton, N.J.: Educational Testing Service, 1962.

Myers, J. L. *Fundamentals of experimental design* (3rd ed.). Boston: Allyn and Bacon, 1979.

O'Connor, G., & Sitkei, E. G. Study of a new frontier in community services: Residential facilities for the developmentally disabled. *Mental Retardation*, 1975, *13*, 35-39.

Pearson, K. On a criterion that a given system of deviations from the probable in the case of a correlated system of variables is such that it can reasonably be supposed to have arisen from random sampling. *Philosophical Magazine*, 1900, *50*, 157-175.

Peters, C. E. *An investigation of the relationship between Jungian psychological type and preferred styles of inquiry.* Unpublished doctoral dissertation, The Ohio State University, 1981.

Rao, C. R. Criteria of estimation in large samples. In *Contributions to statistics.* New York: Pergamon Press, 1965, pp. 345-362.

Reynolds, H. T. *Analysis of nominal data.* Sage University Paper series on Quantitative Applications in the Social Sciences, series no. 07-007. Beverly Hills and London: Sage Publications, 1977.

Stevens, S. S. On the theory of scales of measurement. *Science*, 1946, *103*, 677-680.

Tatsuoka, M. M. *Multivariate analysis: Techniques for educational and psychological research.* New York: John Wiley, 1971.

Theil, H. On the estimation of relationships involving qualitative variables. *American Journal of Sociology*, 1970, *76*, 103-154.

Upton, G. J. G. *The analysis of cross-tabulated data.* New York: John Wiley, 1978.

Wenig, R. G. *Tests of independence for 2 × 2 contingency tables when the sample size is small.* Unpublished masters thesis, The University of Toledo, 1979.

Winer, B. J. *Statistical principles in experimental design* (2nd ed.). New York: McGraw-Hill, 1971.

Yates, F. Contingency tables involving small numbers and the χ^2 test. *Journal of the Royal Statistical Society Supplement,* 1934, *1,* 217–235.

Yule, G. U. On the association of attributes in statistics. *Philosophical Transactions of the Royal Society, Series A.,* 1900, *194,* 257–319.

Zahn, D. A., & Fein, S. B. Large contingency tables with large cell frequencies: A model search algorithm and alternative measures of fit. *Psychological Bulletin,* 1979, *86,* 1189–1200.

INDEX

NAME INDEX

SUBJECT INDEX